Developments in Aquaculture and Fisheries Science, 1

FARMING MARINE ORGANISMS LOW IN THE FOOD CHAIN

Developments in Aquaculture and Fisheries Science, 1

FARMING MARINE ORGANISMS LOW IN THE FOOD CHAIN

A Multidisciplinary Approach to Edible Seaweed, Mussel and Clam Production

P. KORRINGA
Rijksinstituut voor Visserijonderzoek, IJmuiden (The Netherlands)

ELSEVIER SCIENTIFIC PUBLISHING COMPANY
Amsterdam – Oxford – New York 1976

ELSEVIER SCIENTIFIC PUBLISHING COMPANY
335 Jan van Galenstraat
P.O. Box 211, Amsterdam, The Netherlands

AMERICAN ELSEVIER PUBLISHING COMPANY, INC.
52 Vanderbilt Avenue
New York, New York 10017

ISBN 0-444-41332-4

Printed in The Netherlands

PREFACE

"Farming the Sea" is a timely topic. As the Food and Agriculture Organization of the United Nations has repeatedly expounded it is becoming increasingly difficult to adequately feed an ever-growing human population. In many parts of the world, the provision of animal protein is precarious and often deficient. The sea supports a great number of animals rich in protein. Nevertheless, although the sea is vast, and though the total primary production in the sea is of the same order of magnitude as that on land, only about 1% of man's food comes from the sea. Calling on all of our technical know-how, efforts to increase the catches for human consumption have not always led to the desired results. On the contrary, signs of over-exploitation of the sea's natural resources can be seen in many places, and the increase in world catches in recent decades may be predominantly ascribed to harvests of "industrial" fish, destined for reduction to fish meal.

Harvesting only what nature produces freely — and what else does a fisherman do? — is in fact a very primitive way of exploiting a natural resource, a principle abolished on the land thousands of years ago. Growing vegetable crops in monoculture on well-tilled and well-manured soil is the only way to obtain a fair quantity of vegetable food on the land, and animal husbandry in its various forms nowadays brings virtually all of the animal protein of terrestrial origin to our tables. Is it not time to turn from a hunting philosophy to one of farming the sea, as we did so long ago and with so much success on the land? There are some striking cases confirming that this is the way to proceed if we wish to make better use of the sea as a source of food for mankind:

— Experience in Europe shows that areas where oysters thrive will yield a much greater crop after introduction of farming procedures than could be expected from simply harvesting what nature produces when left to itself. A twenty- or thirtyfold increase is not exceptional.

— In northwestern Spain there has been an impressive development of a mussel industry in sheltered waters rich in phytoplankton. Spain is now the world leader in mussel production, whereas some 30 years ago, no mussels fit for human consumption were harvested there.

— For centuries, milkfish have been farmed on the Isle of Java, leading to a crop nature would never produce without man's guidance. In Japan, a variety of marine organisms, from seaweed to shrimp and fish are cultivated with noteworthy success.

In the light of these striking examples it is no small wonder that it is believed that the sea could do much more to feed the human population than it now does with free fishery. Those willing to take up the challenge or to stimulate such a development are in need of factual information. In the first place re the biological aspects, dealing with the environmental requirements of species, their growth and reproduction, and the agencies responsible for the so-called natural mortality: predators, parasites, and diseases included. It is a difficult task to search for such information in the literature. This is not because the written information does not exist, but rather because the data are so widely scattered in publications written in a great number of languages, each publication only dealing with certain facets. Thousands of scientific papers deal with the biology and cultivation of oysters alone. Even the few fisheries biologists who are lucky enough to possess both the annotated bibliography of Baughman (1948) and the supplement produced by Joyce (1972), will find it extremely difficult to grasp the essentials of oyster farming. The farming techniques have to be based on the biological characteristics of the particular species, to be adapted to the local hydrographic conditions, to take the locally available materials into account, and to keep labour costs sufficiently low for the market to absorb the product for at least cost price. Changing social and economic aspects make the older descriptions of the way one once farmed a marine organism no longer applicable in our modern time. Similarly, a description of farming developed in one country will not necessarily be applicable in other parts of the world.

In recent years, several books and scientific articles have appeared on the topics "Aquaculture" and "Farming the Sea", and though such general descriptions are certainly very welcome, they cannot provide all the information needed by someone wanting to set up a farm, by a scientific advisor appointed to assist the farmers, or by an official judging where one could inject some taxpayer's money for the benefit of the country. All these people must feel disappointment in merely finding general descriptions of techniques in use in various parts of the world and the author's considerations at an academic level, often pointing at potential rather than giving a down-to-earth treatment of just how the marine farms really operate from the biological, technical, legal, and economic angles. The fisheries biologist wishes to be properly informed about the environmental requirements of the species and about the farming risks to be ascribed to predators, parasites, diseases and competitors, and, of course, about possible ways of controlling them. The farmer or prospective farmer wishes to be informed on the initial legal steps and on the general hydrographic conditions needed for setting up a farm. He also needs to know the capital investment required, the running expenses recurring year after year, the number of labourers required to produce one ton of the marketable crop, the price fetched at market, and the size an enterprise should be to be remunerative. From a purely biological point of view it may be considered possible to farm almost any organism whose requirements are known, but technical and above all economic aspects govern whether or not remunerative farming is really possible in the area under consideration. The governmental fisheries administrator feels the need of multidisciplinary information before he will decide to stimulate new farming ventures or subsidise existing industries experiencing difficulties. This applies particularly where money has to be poured into enterprises to get a new industry on its feet, and in cases where there is a continuing need for scientific advice to be provided free of charge by the government. Lack of such information will often lead to the sad result of potentialities not being developed.

Multidisciplinary treatises are practically non-existent in this special field. When I was invited to make the effort to write such a book, for the benefit of marine farmers, fisheries biologists, and governmental administrators, I decided to accept the challenge since I have spent most of my scientific career in the field of shellfish farming and have had the privilege to travel quite extensively, paying special attention to a variety of cases of farming the sea, and could make direct contact with farmers in many countries personally, as I can speak, or at least understand, their languages. A multidisciplinary treatment would be a prerequisite and I decided that rather than try to describe all kinds of enterprises in this field all over the world, experimental cases included, I would select a limited number of well-established cases from different links of the food chain and spread over the various climatological zones of the world.

I considered it would be more practical to analyse for each type of farming one well-defined enterprise than to give a description of the industry as a whole. I therefore selected in every area to be dealt with, one company, not necessarily the biggest, but always one which gave the impression of being economically sound and remunerative, and whose pattern was not unduly complicated. That the farmers selected were always willing to provide me with detailed information on various aspects of their enterprise was a pleasant surprise. The consideration that differences in hydrographic, economic and social climate will rarely make it possible to copy their pattern of operations elsewhere no doubt played a part in this attitude. This made it possible to treat all the cases selected in exactly the same multidisciplinary way. To facilitate comparison, each separate case is subdivided in a standard way, as follows:

A. Background
- A.1. General principles
- A.2. Biology of the species
- A.3. Geographical situation
- A.4. Hydrographic pattern
- A.5. Legal aspects

B. Sequence of operations
- B.1.—B.*x*.: This section forms the body of the descriptions giving step-by-step all the required operations from the early developmental phases to marketing, with indications of techniques used and labour involved.

C. Farming risks
 C.1. Hydrographic conditions
 C.2. Predators
 C.3. Parasites and diseases
 C.4. Competitors

D. Economic aspects
 D.1. Rents
 D.2. Inventory items
 D.3. Expendable items
 D.4. Purchase of seed stock
 D.5. Labour
 D.6. Sales

Rather than trying to present a balance sheet holding good for one year only, I decided to give an impression of the sort of expenses the farmers are faced with, of the amount of labour involved, the quantities of materials to be purchased, and of the sales. For the items on the inventory, I give information on the technical aspects rather than on the purchasing price to make it possible for those working in other parts of the world to figure out what a similar item would cost there. Indications of the life of the various items to be purchased, as experienced by the farmers, give a general idea on the investment required. Presenting the economic aspects in such a manner might prevent this book becoming out of date. Many years after its appearance it should still be possible to work out the economic aspects of a certain type of farming on the basis of the prevailing prices.

Every case has sections entitled "Governmental assistance" and "General impression". In the latter I try to formulate my view of that particular case of farming the sea and on its future development. Where needed, a section "Technical details" follows, and every chapter is completed by a "Bibliography".

An effort has been made to reveal the truly dynamic pattern of all cases dealt with. Only when the farmer does his utmost to keep pace with the technical, social, and economic developments in his country can he hope to continue the work which fascinates him so much, which gives work to so many and puts a product on the market which either his own country or neighbouring countries need.

It is felt that the description of the farming techniques applied in various countries under so greatly diverging technical, social, and economic conditions will be very instructive for all those engaged in similar enterprises; the farmers themselves, fisheries biologists who have to advise the farmers, and fisheries administrators in governments. It will certainly offer them various new ideas which may make it possible for them to adapt their operations in a rapidly-changing world.

There are certainly more possibilities of farming the sea than are at present exploited. Unfortunately, it will not be an easy matter to utilise the vast open reaches of the sea, since farming is at least for the time being limited to sheltered coastal areas for technical reasons. Unfortunately, serious impacts are experienced in coastal areas from domestic and industrial pollution, and locally from recreational demands. FAO's ideal to reach a farmed crop production 10 times as high as the present one by the year 2000 seems unrealistic. But farming the sea still offers a multitude of possibilities: the production of food rich in animal protein at a reasonable price level, the production of luxury items for the enjoyment of the epicurean consumer and benefit to the producer.

Since the cases described come from various parts of the world, the information was received in various units. It would, of course, have been possible to convert all length and volume measures to the SI system, but since this would lead to figures after the decimal point, suggesting a precision beyond reality, it was decided to use the units of the country concerned where they are generally understood.

The same applies to the monetary units used: the prices are in every case expressed in the currency of the country concerned. A conversion table fixed at a given date (November 6th, 1972), representative of the period the information was collected, serves to compare the local units with the American Dollar. How quickly these conversion rates may change is indicated by the list for the arbitrary data March 3rd, 1975.

Name	Abbreviation	Conversion rate 6-11-1972	Conversion rate 3-3-1975
French Franc	FFr.	5.035	4.15
Italian Lira	L	584.5	626.5
Japanese Yen	Y	301.0	285.23
Dutch Guilder	Dfl.	3.233	2.337
Philippine Peso	P	6.70	6.87
Portuguese Escudo	Esc.	26.90	23.79
Spanish Peseta	P	63.5	55.46
American Dollar	US $	1.0	1.0
Russian Rouble	R	0.824	0.689
Yugoslavian Dinar	Din.	16.50	15.85

This volume deals with farming of marine organisms low in the food chain. Only one chapter is devoted to farming an organism belonging to the first link of the food chain, a representative of so-called primary production, comparable with the agricultural crops so predominant on the land. It deals with the case of "nori" farming, cultivation of the edible seaweed *Porphyra*. Though this is an industry of considerable economic importance in Japan, it can hardly be ranked with the agricultural crops which provide man's basic food, for "nori" is quite an expensive condiment with limited food value. Farming other species of seaweed is in principle possible and practised on some scale here and there in Japan, both for the production of human food and of material for industrial use. This is more or less still in the experimental phase, so descriptions are not incorporated in this book which deals with commercially-established enterprises.

Marine organisms in the second link of the food chain, herbivorous creatures by definition, are farmed on a large scale. This group comprises filter feeders such as mussels, clams and oysters, and, in principle, herbivorous gastropods such as abelone's (*Haliotis* sp.), and fish which feed on vegetable matter such as mullets (*Mugil* sp.) and milkfish (*Chanos chanos*). This volume largely deals with the various systems of mussel farming practised throughout the world, and concludes with one single case of clam farming. Mussels, feeding on microscopic organisms of the phytoplankton and their remains, can be grown and fattened on a truly commercial scale in well-sheltered waters which are rich enough to support a fair crop of these minute plants. Mussels, well protected against enemies by their calcareous shells, have a meat content of about 25% when in good condition, and the edible parts contain a fair amount of protein as well as a supply of glycogen, the "animal starch" they store as reserve material. Since mussel farming can be mechanised to a certain extent where hydrographic conditions allow, the production costs need not be high, so that mussel meat can easily compete with the low grade meat of terrestrial herbivores produced in animal husbandry. Therefore it can safely be said that mussel farming can substantially contribute to feeding the human population in areas which lend themselves to such an industry. There is no doubt that mussel farming could be practised in several other countries favoured with a well-indented coastline bathed by water rich in phytoplankton. Still, one should be cautious of establishing new industries, for one of the serious constraints to mussel farming is poor consumer acceptance. In several countries mussels are a very popular sea food, presented in highly attractive main dishes such as "moules marinières" or as an appetizer such as "moules à l'escargot", but in other countries the demand may be virtually non-existent. This is partly due to conservative feeding habits, partly to lack of experience in eating a high-quality farmed product and partly, no doubt, to a wide-spread aversion to invertebrates as food. The most surprising case is Japan, so dependent and so keen on sea food such as oysters and various clams, where mussels could easily be farmed on a large scale at low cost, but where the general consumer considers mussels repulsive! It will surely take a lot of time and persuasion to alter such unfounded aversions. If the poor consumer acceptance is based on disappointing experience with the product from natural beds, often poor in quality and sometimes containing many small pearls, things may change more rapidly when high-quality farmed mussels become available.

Another aspect of mussel farming, very recent in nature, is the possible use of mussels in the pharmaceutical industry. If the first medical experiences are corroborated this might eventually lead to an unexpected stimulation of mussel farming. The very recent onset of mussel farming in hanging culture in New Zealand, based on the species *Perna canaliculus*, is due to the new pharmaceutical interest.

That this volume winds up with only one chapter dealing with clam farming has a sound reason. Though clams are a highly estimated food in many countries, most of them are just harvested from the natural beds. The case described, the farming of little-neck clams in Portugal, deals with a well-established industry. In Japan farming of closely-related species like *Venerupis philippinarum* and *Meretrix meretrix* is practised on some scale, but merely consists of transplantation, to avoid overpopulation and to promote growth. A very recent development is scallop farming in northern Japan (Hokkaido), where a system of spat collection in the species *Pecten yessoensis* exists. This industry is still too young to be described as an established practice.

The future has perhaps new possibilities in store, for it is now possible to produce any quantity of small seedlings of various clams of the genera *Venerupis, Meretrix, Mercenaria, Arca, Pecten*, etc., and also of the herbivorous gastropod *Haliotis*, in a hatchery. There is still a difficulty to be overcome: how to grow the tiny and vulnerable seedlings on a large scale with minimal losses to a size large enough to withstand the attacks of the many predators occurring under natural conditions.

The first chapter of this volume entitled "Introduction", gives the scientific background and basic philosophy for the entire field of farming marine organisms, dealing not only with creatures of the first and second link of the food chain, but also with those of higher links, the carnivorous species. This anticipates the appearance of the following volumes dealing with farming of oysters, fishes, and shrimps. I decided somewhat arbitrarily to incorporate in these volumes all those "marine" organisms which reproduce in the sea but may sojourn in a later phase of their life in brackish or even fresh water, and also other organisms which are farmed for the greater part of their life in sea water.

Professor Dr P. Korringa

ACKNOWLEDGEMENTS

It would have been impossible to write this book without the willingness of all the managers and owners of the enterprises chosen as examples to inform me in such great detail about the technical and economic aspects of their work. Few of these matters have ever been described together before, but all the farmers with whom I discussed their enterprises understood the advantage in giving access to their long experience in this field. I feel greatly indebted to them all for the frank and friendly way they replied in such great detail to the many questions I put to them.

For the hydrographic and biological background, I was very kindly assisted by colleagues in many countries. I am very grateful for the patience with which they formulated replies to the many questions I had to put to them to reach a full understanding of the local situation. I would like to mention a few people in particular: Mr P. Businger, Agricultural Attache of the Royal Netherlands Embassy at Tokyo, saw to the collection of detailed information on "nori" farming in Japan, assisted by his secretary Y. Akimoto. They paid a visit to the enterprise selected, took photographs to elucidate the case and had the patience to write elaborate replies to my long questionnaire.

For the chapter on mussel farming in the Netherlands it was Mr A.C. Drinkwaard, Head of the Mussel Experimental Station on the Isle of Texel, and Mr L. Westbroek, both belonging to my institute, who provided me with technical and hydrographic information and with several photos showing details of the industry.

Mme Y. Dardignac of the Laboratory of the "Institut Scientifique et Technique des Pêches Maritimes" at La Rochelle kindly assisted me in collecting additional information on mussel farming in the Charente Maritime.

For mussel farming in the Ría de Arosa, Dr B. Andreu and Dr A. Figueras of the "Instituto de Investigaciones Pesqueras" respectively at Barcelona and at Vigo, and Dr R. Robles of the "Instituto Español de Oceanografia" at Vigo supplied me with valuable additional scientific information.

In the case of mussel farming near La Spezia (Italy) it was Dr^{ssa} G. Nacini from Lerici who put not only her entire collection of photos dealing with the local mussel industry at my disposal, but also her manuscript on the economics of mussel farming in that district.

Mrs M. Hrs-Brenko of the Centre for Marine Research in Rovinj was kind enough to provide me with a lot of additional information on mussel farming on the coast of Istria, and Mr Z. Filić and Mr G. Sošić from the same institute made technical drawings and photos for me.

I am very much indebted to Dr A. Ivanov of AzCherNIRO at Kerch for sending me so many detailed letters, despite the language difficulties, about his experimental mussel farming. It would have been impossible to gain a real understanding without his kind and patient assistance.

Mr I. Ronquillo of the Bureau of Fisheries at Manila, aided by the fisheries biologist R.N. Pagcatipunan, provided me with detailed information on the young mussel industry in Bacoor Bay, The Philippines.

My old friend Dr H. Vilela, former director of the "Estaçao de Biologia Maritima" at Lisbon gave me the detailed verbal and written information I requested on farming of little-neck clams in Portugal, and also offered me some of his pictures on the subject.

Mr B. Verboom of my own institute kindly consented to make a great number of drawings and charts for my book to ensure presentation and good elucidation of technical details.

Last but not least, I wish to express my gratitude to my industrious and conscientious secretary Miss E.L. van der Weele, who kept the wealth of material in order and who typed so many letters and the entire manuscript.

CONTENTS

INTRODUCTION

Farming, in the widest sense of the word, made it possible for human beings to live on this world in such enormous numbers. Agriculture and livestock breeding form in reality the material roots of our civilisation. In the background of all the daily consumption of food by mankind stands the biochemical reaction of carbon dioxide assimilation, of which green plants still hold a monopoly. Green plants of various description build up organic constituents such as carbohydrates, proteins, fats and vitamins, with the aid of energy derived from the sunlight, from the basic constituents water (in liquid form), gaseous carbon dioxide, and the so-called nutrient salts, consisting of a variety of soluble inorganic constituents containing all the elements a plant needs for its growth and reproduction. Vegetable matter largely prevails in man's diet, but all foodstuffs of animal origin are primarily built up from the organic material which the green plants have produced with the aid of sunlight.

On land, herbs, shrubs and trees predominantly do this job, and algae and mosses to a lesser degree. Agronomists have measured how much organic matter the plants produce per unit time and surface area. On well-tilled and manured land one may expect a production of 200 to 1000 g of dry organic matter per m^2 per year, this calculation covering all climates and various crops. It is clear, however, that one cannot expect a crop of this productivity on every m^2 of the different continents, for one should take into consideration that the factors sunlight, water (in liquid form) and nutrients, occasionally also carbon dioxide, may be locally so scantily available that they limit the production. Deserts, steppes, high mountains, and ice-covered areas fall far short of the production figures given above.

It is not surprising that virtually the same factors are involved in primary production in the aquatic environment, but a reliable quantitative understanding of this matter was only obtained when the ^{14}C method had made it possible to measure the photosynthetic performance of the phytoplankton in water samples taken from different depths. It soon became clear that seaweeds and eelgrasses, locally found in dense vegetation along the coast, contribute only a small fraction of the organic matter produced in the sea, whereas the representatives of the phytoplankton, organisms of microscopic dimensions belonging to various taxonomic groups, take care of almost the entire crop of organic matter produced by photosynthesis in the aquatic environment. The same basic factors determine the quantities produced per unit time and surface area, but contrary to the situation on land, the factors water (in liquid form) and carbon dioxide are never the limiting factors in the sea. Light and nutrients form the limits to primary production. Recent investigations revealed that about 1/3 of the ocean's surface layer should be considered as a desert, characterised by a deep blue colour of the water, producing less, often far less, than 100 g of dry organic matter per m^2 per year, and less than 1/10 of the ocean should be reckoned as being fertile water, producing 500 to 1000 g/m^2 of dry organic matter per year. Considering that the ocean covers about 71% of the surface of the earth, one is forced to conclude that the total production of organic matter by the photosynthetic activity of green plants must be of the same order of magnitude on land and in the ocean. However true this may be, one is faced with the unmistakable fact that only about 1% of all man's food comes from the sea!

THE FIRST LINK IN THE FOOD CHAIN

How should one explain this remarkable discrepancy? Primarily by realising that about 85% of the crops harvested on the land consist of vegetable matter and only 15% of products of animal origin, such as meat, milk and eggs, whereas the vegetable crops harvested from the sea constitute only a minor fraction of man's sea food. Man harvests or farms some seaweeds for direct consumption, predominantly species of the red-weed *Porphyra*, but since the total crop harvested globally should be counted in tens of thousands of metric tons dry matter per year, seaweeds contribute only a few grams per year to man's overall diet. Contrary to the terrestrial situation, the reasons why such a minor fraction is taken from the first link of the food chain in the marine environment should be sought in differences in density, dimensions and diversity of the vegetation. On land the vegetation is predominantly concentrated in a layer about 1 m thick, partly in the soil itself, partly in the air, whereas in the water the assimilating phytoplankton is dispersed throughout a much thicker layer, down to a depth of 100 m, in areas where the water is very clear. Plants and parts of plants harvested as human food on the land are of dimensions which allow them to be collected easily by hand or machine, whereas in the water only the seaweeds have these properties, and the organisms of the phytoplankton are so small (from 0.01 to 0.5 mm in diameter) that harvesting by hand is out of the question and mechanical harvesting on a commercial scale extremely difficult and costly. The vegetable crop on the land is predominantly grown in monoculture, removing all unwanted species as "weeds", either by mechanical or chemical means. Only on a very small scale does one pick edible plants or parts thereof from a wild mixed population, whereas in the sea one finds just such a mixed population of phytoplankton, varying in composition in space and time. Many phytoplankton organisms have a siliceous frustule, often adorned with long spines, and some others are either poisonous or repugnant to the taste. It is clear that such a mixed crop could not serve as food for man or his cattle, even if harvesting became technically possible. Selection of palatable and digestible items from such a natural assortment of microscopic organisms is therefore out of the question.

THE SECOND LINK IN THE FOOD CHAIN

About 15% of man's food produced on the land is of animal origin, predominantly meat, milk and the milk products butter and cheese, and eggs. Upon closer inspection one realises that this 15% originates from the second link in the food chain, from animals which are herbivorous or at least predominantly so. They feed upon vegetable matter and convert part of it into their own tissues, popularly known as meat, milk, eggs, internal organs, skin and bones. It is, however, only part of the food which has this destination, the major part being used for the daily metabolic requirements such as energy for locomotion, for keeping the body temperature at the desired level, for the digestive processes, etc. Which part of the food will be converted into animal tissues or into food reserves for the next generation (eggs) depends on the type of food, the developmental stage of the animal, and the conditions under which it lives. Under conditions of animal husbandry by modern methods the animals need make little effort to collect their food, and will be duly protected against the hardships of the climate. Therefore their conversion rate will be quite different from that in animals living in the wild, which have to cover considerable distances in their daily or seasonal feeding migrations, spending a lot of energy in efforts to escape from their predators, and having to spend their whole life in the open air. As a reasonable average one assumes that about 10% of the food ingested will be converted into animal tissues, milk included; this fraction is greater under husbandry conditions, occasionally considerably greater, but smaller in the wild, and considerably smaller in older animals. If 10% represents the average conversion rate, one must conclude that the 15% food of animal origin which man consumes world-wide, requires more vegetable matter than man consumes himself directly from the first link of the food chain. This explains why foods of animal origin are more expensive than vegetable foods. Nevertheless, man wishes to include animal product in his diet because it is palatable and easy to digest, and because it contains a balance of amino acids similar to that in man's own tissues. Were it not that most of the herbivores used in animal husbandry are capable of digesting foodstuff such as grass, indigestible for man, the price of animal products would be considerably higher than they already are. Upgrading vegetable matter into animal products does increase the quality of the food, but inevitably leads to a noteworthy decrease in quantity.

If such is the situation on the land, where herbivores are exploited by husbandry methods to satisfy man's need for animal products, a higher percentage of which he uses when he can afford to, how is the situation in the sea? Since, for reasons explained above, the first link in the food chain produces but a dwindling percentage of the food man collects from the sea, does one perhaps make use of herbivorous species as the food product of the sea? To a certain extent one does, especially in the form of the so-called filter feeders. Phytoplankton is food for a variety of animals of different taxonomic groups, many of which belong to the zooplankton. Most of the zooplankton organisms are of such small dimensions that man cannot use them directly as food. Even if they were found to be palatable, it would be extremely difficult technically to collect them selectively on a large scale, and also to eliminate the indigestible parts, such as chitinous carapaces or calcareous shells. There are, however, larger invertebrates which feed on phytoplankton, or their remains, by filtering them from impressive volumes of water. Evidently this requires a lot of energy, so that the conversion rate cannot be very high, but when the phytoplankton is rich enough and consists for the larger part of ingestible and digestible species, such filter feeders may grow and fatten rapidly.

Since prehistoric times man has recognised such marine filter feeders as wholesome and palatable food, and has collected them on a large scale, predominantly in intertidal areas. Shells of oysters, mussels, cockles, scallops and clams found in many prehistoric kitchen middens bear witness to this. Bivalve molluscs constitute by far the greater part of all the creatures of the second link in the food chain that man takes from the sea. In addition, he utilizes some browsing gastropod molluscs such as periwinkles, which also belong to the second link in the food chain, and locally he even takes creatures such as Ascidians. Though the total quantity of mulluscan shellfish used as human food is quite important when expressed in tons weight, it constitutes only a minor fraction of all the food man takes from the sea. Among the marine vertebrates there are a few species with herbivorous habits, feeding on the first link in the food chain, and they too are incorporated in man's diet: fish such as the mullet (*Mugil* sp.) and milkfish (*Chanos chanos*) are of some local interest as contributors to man's diet, but the world production of these species is quite modest.

THE HIGHER LINKS IN THE FOOD CHAIN

The animal-based food man produces on the land virtually all belongs to the second link in the food chain. Carnivores are rarely eaten by man, not because they would not be palatable, but rather because the number of carnivores occurring in nature per unit of land area is so much smaller, due also to the 10% rule, than the number of herbivores, and because the husbandry of such species requiring meat would result in prohibitive prices for the end products. One could point out that quite cheap animal proteins such as fish meal are used in the diet of chickens, ducks and pigs, albeit for economic reasons, on a limited scale only. Hence, we should not consider these end products as belonging purely to the third link in the food chain, on the same level as hedgehogs, opossums, woodcocks and snipes, which are occasionally admitted to the table, although extremely small quantities of such foodstuffs are available.

How different is the composition of the food man takes from the sea! Close to 99% of all the organisms man collects from the sea to serve him as food belong to the third, fourth or even fifth links in the food chain. Since the passage of food from each link to the next in the food chain means at least 90% loss (under natural conditions), it is clear that only limited quantities of foodstuffs can be taken from the higher links, even where the primary production by organisms of the first link in the food chain is quite high. This habit of man of collecting predominantly the animals of higher links in the food chain from the aquatic environment, explains why only 1% of man's food comes from the sea, although the primary production is of the same order of magnitude in the sea as it is on the land.

THE FISHERY

It is the fishing industry which collects the edible denizens from the sea. Fortunately, the species belonging to the higher links in the food chain are not evenly distributed throughout the waters of the oceans, but congregate markedly (often seasonally) for several reasons, which makes possible remunerative local fishing. That about half of all the food man takes from the sea comes from beneath less than one tenth of one percent of the surface of the sea, is, however, surprising! This truly remarkable fact should not be ascribed to man's inability to go and fish wherever fish is to be found. In olden times man, for reasons of safety, simply had to fish in the coastal waters, but in our time ship-building, fishing-gear construction, navigational aids and electronic fish-finding equipment have reached such a state of sophistication that it is possible to fish wherever one wishes. That efforts are still focused on very localized parts of the sea is explained in terms of fish concentration. Only where the primary production is very high, can one expect a greater biomass of organisms belonging to the higher links in the food chain. Therefore in the fertile regions at higher latitudes, and in the areas with upwelling water which one encounters here and there, the most promising rewards for the fishing industry are to be found. In the tropical and subtropical regions of the sea the surface layer, warmed by the sun, and thus of lower specific gravity , will not mix with the colder, deeper layers of the sea at any time during the year. Only where the sunlight penetrates with sufficient intensity can organic matter be produced by the phytoplankton, the organisms of the first link in the food chain, provided that nutrient salts are available in adequate quantities. The warmer upper layers of the ocean become, however, increasingly poorer in elements of vital importance to the phytoplankton. Dead matter, faecal pellets and exuviae tend to sink to the deeper layers, and biodegradation by microorganisms is not sufficiently rapid to release all the nutrients into solution before a depth is reached where sunlight is too weak for carbon dioxide assimilation. That is why in the tropical and subtropical zone the surface layers are, over large areas, poor, and why upwelling of water from great depth, cold but rich in nitrogen and phosphorus compounds, leads to a high productivity. A second reason for the uneven distribution of the fish stocks is the fact that most of the fish collected as man's food need to live either permanently or temporarily in shallow water, either because they collect their food on the sea bed and only where the water is rich in plankton throughout the range of its depth can one expect a rich thriving stock of benthic creatures, or because they need shells, pebbles or rocks in shallow sites on which to deposit their eggs. Hence, most of the rich fishing areas are to be found on the continental shelves, especially at higher latitudes, and in areas where upwelling occurs regularly. All the seas together nowadays yield 60 to 70 million tons of fish and shellfish per year, partly used as human food, and partly as raw material to be reduced to fish meal and fish oil and only indirectly used as human food. Recent increments in the landings are largely due to an increase in fishing for industrial purposes. A rich sea like the North Sea produces some 3 million tons of fish per year, about 1/20 of the world production, though this sea comprises only 1/25000 of all the water contained in the oceans.

Efforts are made to increase the yield of fish — preferably edible fish — from the sea, but in most of the productive areas one observes already the unmistakable signs of over-exploitation, such as a reduction in the average size (age) of the fish, a decline in the number of year-classes caught together and therefore noticeable fluctuations in the annual catches, this being detrimental to the fish trade, and a decline in the catches per unit of effort. International legislation would make it possible to improve the situation and to approach the optimum sustainable yield, or if preferred, the optimum sustainable profit, but since many countries participate in the fishing industry this is extremely difficult, and improvement will be a very slow process. Especially where fish fit for human consumption, such as herring and mackerel are used on a large scale for reduction to fish meal, as in the North Sea, agreement on the measures to be taken to improve the situation is hard to reach. In reality fishing should be considered as a rather primitive way to harvest the bounty of the sea.

FARMING THE SEA

If one considers that in civilised regions the system of harvesting land crops produced freely by nature was abandoned thousands of years ago by switching over to agriculture and animal husbandry, one wonders whether it is not high time the same thing was achieved in the sea, and whether this could not lead to a much higher production than the modest 1% man now obtains from it. This would mean reducing the area allowed for free fishery and putting a greater area into the hands of prospective farmers of the sea, who would aim to reduce drastically the number of species in such areas in an effort to grow a crop in monoculture. Adequate control of predators and competitors would ensure that a reasonable percentage of the local primary product would ultimately be converted into the substance of the desired species. Inevitably such a procedure would involve a considerable investment of labour and money. Is such a procedure possible in the sea? It certainly is, and it is easy to adduce a few examples to show that in the sea, farming can also lead to a far greater production of human food than free fishing:

(1) Exactly 100 years ago the Dutch Government withdrew freedom to fish on the natural oyster beds in the Oosterschelde, and passed the grounds into the hands of prospective oyster farmers. Appropriate farming techniques (including control of predators, competitors and diseases), supported by scientific research financed by the government, led to a spectacular rise in the production of oysters. The original natural beds yielded only about half a million oysters per year, but by the adoption of farming techniques, up to 30 million marketable oysters were produced in the same waters year after year. Evidently nature, left to itself, will not necessarily produce the optimum quantity of oysters for man's purposes.

(2) About 25 years ago exploitation of the rich phytoplankton in the "ría's" of Galicia (N.W. Spain) was begun by growing mussels in hanging culture. In due course one learned to design special rafts and to suspend from these the ropes carrying the mussels. Within 14 months the tiny mussel seed, collected from the rocks in the intertidal zone, grows to become a large-sized marketable mussel. Now this area produces almost 160 million kg of marketable mussels per year, whereas nature, without man's intervention, produced no mussels here fit for human consumption.

(3) Hundreds of years ago shallow ponds were constructed on the North coast of Java (Indonesia) and into these ponds, which were filled with a mixture of sea water and fresh water, juvenile specimens of the milkfish (*Chanos chanos*) were introduced. The milkfish is a marine species which grows and fattens best in brackish waters with a rich vegetation of blue-green algae, on which it feeds. The milkfish, therefore, is one of the few fish belonging to the second link in the food chain. Therefore one can expect a particularly high production of edible fish per unit of surface area. Nowadays thousands of milkfish ponds are in use in Java and Madura, but also in the Philippines, in Taiwan, and in some other countries in south-east Asia. This form of farming the sea has certainly led to a far greater production of milkfish than that found under natural conditions.

(4) In Japan similar ponds have recently been constructed to cultivate

another marine fish, the eel. The elvers, born at great depth in the ocean, travel to inshore waters and try to reach low salinity waters rich in food. When approaching maturity they begin the return journey to their spawning grounds in the depths of the ocean. The eel ponds do not contain enough food to allow a greater number of eels — which belong to the third and fourth links in the food chain — to grow and fatten, but Japanese eel farmers, guided by scientific advice, have in recent years learned how to feed the eels with a formulated food, and how to control the quality of the water in the ponds, so that the production of marketable eels has now reached an almost incredibly high level per unit of surface area.

It is not difficult to list many more examples of farming the sea: the oyster industries in France, Australia, Japan and the United States are among the more important economically.

Do the cases mentioned suffice to warrant the conclusion that farming the sea by the development of modern techniques will in general lead to a much higher production of human food from the marine environment than the traditional free fishery? Will a more general adoption of farming techniques lead in due course to a more reasonable contribution towards human food production from the saline waters, than the humble 1% one has to record now?

The answer must be that those cases of farming the sea mentioned above are fundamentally not comparable with the traditional systems of agriculture, horticulture and animal husbandry. It is true that one works on privately owned grounds, that one protects the crops against enemies of various description, human intruders included, that one has learned to lend a helping hand in overcoming the various constraints which act upon the species concerned. But is all this really the same as what we call "farming" on the land?

Let us look at the case of oyster farming in The Netherlands. The production from the original natural beds was limited by a scarcity of objects ("collectors") on which the planktonic oyster larvae could settle. Hence the great success when man offered the larvae suitable collectors at the appropriate moment. A second point is that the settlement of oyster spat can be expected in places where current velocities are low for hours in succession. But growth and fattening are invariably much better in places where somewhat stronger currents, richer in phytoplankton, prevail. Hence the success obtained by the transplantation of seed-oysters to such places. But is this a case of real farming or rather of semi-farming? The Dutch oystermen put out their collectors, but then must wait and see whether water temperature, availability of food, and absence of predators will favour the development of the oyster larvae, and hence they depend entirely on nature for the success of their spat-producing efforts. If the oystermen do manage to produce a great quantity of oyster spat and then try to step up the production of marketable oysters, they will find, and in fact have found repeatedly, that the Oosterschelde will never produce more than the 30 million oysters mentioned earlier. As soon as the local oyster population exceeds a certain level, invariably growth slows down and mortality strikes unusually hard. Evidently the amount of phytoplankton in the water of the Oosterschelde puts a limit on the production of oysters. Nature sets the upper boundary; man can do little to offer the oysters more food than nature itself produces. Investment of more capital will not lead to a greater production in this example. In a modern chicken factory or in a mushroom farm, on the other hand, the grower decides himself how large production will be, and need not wait and see whether nature will be generous or not. The oyster farmers cannot hope to exceed a given level of oyster production; they have to accept ups and downs in spat production commensurate with fluctuations in larval development, and regard it as normal to see a given percentage of their oysters die each year; they cannot avoid heavy losses brought about by severe winters, gales and other natural phenomena. Oyster farming (Dutch style) must be defined, therefore, as a case of semi-culture.

Mussel farming in Galicia has undergone a spectacular development and reveals several of the features of a real industry. The quantities of mussels produced are very great indeed, every rope carrying about 50 kg of mussels, with hundreds of ropes hanging from each raft. The Spanish mussels are produced in hanging culture, whereas in The Netherlands, mussels are grown on culture beds, a good plot yielding about 10 kg of mussels per m^2. It is often considered that this is a very high concentration of animals, even though the edible parts constitute only about 25% of the mussel weight. One should not however compare the 10 kg of mussels grown on 1 m^2 of sea bed with the number of cows on a pasture, or sheep on a moor. The difference is that the mussels do not live in stagnant water and are therefore not dependent on the standing crop of phytoplankton in the water column immediately above them. The tides bring phytoplankton from a vast area, and calculations on the production of animal matter should be based on that area and not on the area of the beds on which the mussels are actually grown. The same argument holds good for the mussels grown in suspended culture in Galicia. Food comes from a far greater volume of water than that in which the mussels are bathed. Though the ría's of Galicia are rich in phytoplankton, and though water temperatures are ideal for mussels all the year round, mussel production cannot be increased indefinitely.

One feels forced to conclude that in this case too, man cannot break the chains nature has set to limit production. Therefore, one should also speak in this case of a semi-culture.

What about milkfish farming on Java? Milkfish feeds on blue-green algae and therefore is of the second link, just like the oyster and the mussel. Hence the possibility of obtaining a large production of fish per unit of surface area. Indonesian fisheries biologists know how to stimulate the growth of blue-green algae with a well-chosen mixture of fertilisers, and therefore it seems as if milkfish farming is dependent to a lesser degree than shellfish farming on the primary production provided by nature. This may be true, but the main problem in milkfish farming is not the quantity of food in the ponds but rather an increasing scarcity of young fry. The milkfish spawns at some distance from the coast and the young fry are caught when they approach the coast in search of brackish water. The factors governing the amount of milkfish fry are only poorly known. Pond-reared milkfish do not reproduce in brackish waters and efforts to speed up the ripening of the ovaries by means of hormone injections have not yet led to practical results. Only when artificial maturation, spawning and fertilization become routine, can milkfish farming be referred to as a case of true farming. One could then control both the extent of reproduction and the production of food, so that if a good system of controlling competitors, predators, parasites and diseases also existed, one could aim at a production of a given magnitude, practically independent of events in nature.

The same holds good for eel farming in Japan. Though the eel belongs to the third link in the food chain, it is possible to grow considerable quantities per unit of surface area because there is now a good formulated food available, and because diseases and parasites may be controlled, and a satisfactory oxygen level in the water of the ponds may be maintained. But just as in milkfish farming, the chief problem in eel farming is procuring the fry. Even if one has a sufficient pond area, all the equipment required, as well as sufficient capital for buying a formulated food, electricity and labour, one cannot hope for commercial success unless elvers are available in sufficient numbers to start the production of a new crop. In years when elvers are scarce, Japanese eel farmers buy elvers in other countries, recently even in Europe, and do not hesitate to pay the high price asked by European fishermen for elvers intended for direct human consumption! Therefore Japanese eel farming is just like the milkfish farming, in reality only a semi-culture, for at least part of the process remains at the mercy of the caprices of nature. Control throughout the life-cycle, so characteristic of a true culture, is not yet possible.

PREREQUISITES FOR WHOLE LIFE-CYCLE FARMING

It will be clear from the foregoing that the development of true cultivation systems for various species of fish and shellfish requires effective control of two important factors: procuring the fry and providing the food. In the case of molluscan shellfish, pioneers such as Cole, Imai and Loosanoff worked for many years on a system of rearing the larvae which were obtained by induced spawning. Success was finally obtained for several species of bivalves and gastropods; a very high percentage of the larvae can be reared up to the benthic phase and beyond, to full size and maturity. The first phase, larval development, requires meticulous care and the utmost cleanliness. Consistent results are only obtained if food-organisms are grown in pure culture and are offered in the correct quantities. Even slight deviations from the scientific recommendations can lead to repeated failure. It is still a specialist's job to produce the desired quantities of young spat, but even the specialist cannot always guarantee prompt delivery of an ordered amount of spat. Neither have efforts to avoid the dangers of a dense monoculture by using a more extensive rearing technique given truly reliable results.

After settling, the young molluscs should be provided with large quantities of running sea water rich in natural food, for at that stage offering laboratory-reared phytoplankton would be financially prohibitive. Provision of sufficient natural food and protection against natural predators and other enemies is, however, still a very difficult combination of requirements. Millions of molluscs belonging to various species have already been cultivated from egg to egg in a full cycle, especially in the United States, in Great Britain and in Japan, but despite many sweeping statements, a commercially sound and independent industry based on these achievements has still to be developed.

Though it is still uncertain whether hatcheries will ever play an important role in providing shellfish farms with the seedstock they need on a truly commercial scale, one should realise that the development of these techniques may prove to be of the greatest importance for the future of these industries. They open up the possibility of selective breeding, recognised as an essential part of agriculture and lifestock husbandry, but thus far virtually unknown in shellfish farming. In principle it ought to be possible to breed strains able to reproduce at given water temperatures, strains able to withstand low temperatures, strains which grow rapidly, strains resistant to certain diseases, or strains with a more attractive organoleptic quality. Since the generation time will virtually always be more than one year, selective breeding will no doubt be a long term venture, and therefore it will be carried out preferably in governmental installations.

Rearing larvae from naturally or artificially ripened parent stocks of fish and shellfish certainly is a prerequisite for full-cycle farming. But, in addition, one should also be able to stimulate the primary food production in order to exercise the necessary control over the quantity of food required in rearing the young specimens as far as marketable size. Nutrients are often the limiting factor in the primary production. During the Second World War experiments with fertilizers were carried out in two Scottish lochs. Though it appeared to be possible to stimulate the primary production in the lochs with superphosphate, one could not confine the effect to those species of plants and animals which are links in the food chain that includes fish for human consumption. Almost all of the fertilizer was ultimately taken up by a variety of plants and animals considered to be "weed" organisms, birds included. This is not surprising, for it is as if one were trying to produce more lions on the African steppe by spreading fertilizer. The food chain, grass—antilopes—lions, certainly exists, but the casual spreading of fertilizer does not guarantee that all the fertilizer used, or even a reasonable part of it, will ultimately be found in the bodies of the lions. If one cannot control the system by eradicating "weed" organisms in every link in the food chain, as is the practice in agriculture and animal husbandry, there is little hope that fertilization of sea water will pay for itself, or that consistent results will be obtained.

There are some cases which can be classified as truly full-cycle farming, the most striking examples of which are "nori" culture in Japan, described in detail in the section dealing with farming within the first link in the food chain, and farming salmon in the sea as carried out in Norway. The latter example is described fully in the last chapter of the volume on fish farming. Cases such as the farming of Penaeid shrimps come very close to the pattern of full-cycle farming.

PROSPECTS OF FARMING THE SEA

The industries based on a system of semi-cultivation such as mussel farming and oyster culture are certainly of considerable commercial importance, and their products are greatly valued on the market, where they fetch good prices. Farming of molluscan shellfish is however confined to coastal waters where there is sufficient protection against gales and other extreme weather conditions. When special equipment such as racks and rafts are used, a fair degree of shelter may be of greater necessity than for cultivation on the sea bed. Suitable grounds should be closed to free fishing and passed into the hands of shellfish farmers or to companies or cooperatives set up for the purpose of shellfish farming. Adequate safeguards against poaching and trespassing should be imposed. Since some phases of farming oysters, mussels and clams of various description require a considerable amount of manual labour, and since a high degree of mechanisation is not always applicable, this type of farming can best be carried out in well-populated areas. Since the cultivated product is perishable, and since there may be a preference to eat it in the raw state, a densely-populated area in the immediate vicinity, or alternatively, excellent transport facilities, are important factors. On the other hand, a densely-populated area usually means increased pollution in the inshore waters, especially where industry is prevalent. There is moreover a growing interest in areas for recreation. In the vicinity of large cities the authorities concerned will not be free to reserve considerable areas for farming the sea. The sad result is a serious clash of interests. Many formerly productive oyster beds are already so badly polluted that larval development and settling are impossible, and conditions for growth and fattening of oysters have deteriorated so much that the oysters have disappeared completely. In cases of less serious pollution the shellfish produced are often tainted to such an extent that they are not accepted on the market or are rejected by public health authorities because of an intolerable level of contamination with domestic sewage. Not only in Europe and the United States, but also in Japan, oyster farming is locally ousted out of the well-protected bays and inlets by an ever-increasing water pollution, and is forced into less-sheltered sites where the risk of damage by severe weather and by predators is greater. A similar clash must be expected where construction works cut off bays and estuaries and consequently lead to a deterioration of the biological environment; this means destruction of existing and potential resources in the coastal waters. Some countries with a well-indented coastline and coastal waters rich in phytoplankton still offer great possibilities for setting up or elaborating a system of shellfish farming, provided that legal aspects, labour, and transport facilities are adequate. One must realise however that the ever-increasing human population tends to increase the pollution of coastal waters, and demands greater areas for recreation, the very areas to which shellfish farming is confined. Therefore the general prospect is that many of the traditional shellfish farming areas will little by little loose their suitability, unless serious efforts are made to control water pollution.

Since molluscan shellfish and some fish such as milkfish and mullet belong to the second link in the food chain, systems of farming based on the natural production of food have a fair chance of being remunerative.

On the other hand, for species considered to belong to the third or fourth links in the food chain, one cannot hope to develop a system of intensive farming based on the natural production of phytoplankton. Farming of such species either as semi-culture, starting with the young fry collected from the natural population, or as true or full-cycle farming, is in principle possible. In the latter case too, well-sheltered ice-free inshore water, sufficiently free from pollution, is a prerequisite for successful farming, since either ponds or netting enclosures must be constructed in order to prevent losses by migration and predation. Since rapid growth is necessary for remunerative fish farming, tropical and sub-tropical countries have an advantage over countries with prolonged periods of low water temperatures. Locally, the cooling water from power plants might be used for heating the sea water for the fish farm. Commercially sound systems of fish farming are based moreover on rapidly growing species, which can be ready for the market within 10 months or less. Considering the care required to grow fish and crustaceans to the marketable size, and the quantities of high quality food required to achieve this growth, one is forced to conclude that the products of such sea farming must necessarily fetch a very high price on the market. In other words, the farming of carnivorous fish and shrimps is certainly possible, but the end-products should be included among the luxury foods. In rich countries there is certainly a market for expensive high quality sea-food, and developing countries could use such products for export to the richer countries, and buy in return at least tenfold the weight of a product such as salted low grade fish to meet the protein needs of their own people.

A biologist will state that in principle every species of plant or animal can be farmed, provided that one is so well-informed about its biology that one can offer the exact space required, the appropriate climatological and hydrographic conditions, adequate food in well-controlled quantities, and that one can control its predators, parasites and diseases. In practice, however, one is only interested in farming if it is economically sound. For species of which considerable quantities are also caught in the wild, one should expect a price for the marketable product not higher than the fisherman gets for his catch. This limits the farming possibilities drastically.

Farming the sea is certainly a realistic possibility in many countries both for second-link products such as molluscan shellfish, mullet and milkfish, and for luxury quality fish and crustaceans from the third and fourth links in the food chain. Experienced farmers, guided by biological and economic advice, will certainly be able to make money this way. Second-link products may be farmed in large quantities in areas rich in organisms of the first link in the food chain. The quantities produced may contribute significantly to the food requirement of the national population, and may provide a higher percentage of animal protein in the daily diet. For the time being molluscan shellfish offer the best possibilities of reaching this goal, though industrial pollution, increasing wages, recreational demands, and catastrophic mass-mortalities of the shellfish, certainly stand in the way of a dramatic increase in production. From the moment a good technique of producing the fry of fish such as the mullet and the milkfish is developed, a large increase in the tonnages produced in different parts of the world is to be expected.

An important factor to take into consideration in initiating the farming of molluscan shellfish is whether there is a market for the product, for in many countries conservative eating habits, prejudices and more or less vague religious precepts still stand in the way of large-scale utilization of molluscan shellfish as human food. But even the tasty mullet is not everywhere held in high esteem.

One should realise on the other hand that farming organisms from the higher links in the food chain may be commercially interesting both from the view-point of the epicurean consumer and the farmer, but the quantities brought to the market will always be modest, for even when the rearing techniques are perfect, the amount of protein-containing feed required must almost always be high. This not only means high costs for feeding, and a high price on the market which only a few can afford, but also the withdrawal of an appreciable quantity of potentially valuable animal matter from the world market. Farming chickens and pigs on a diet consisting partly of fish meal means that a considerable tonnage of fish such as herring, mackerel and gadoids never arrives as such on the food market, but appears there in some upgraded form, higher in quality and in price, but very much less in quantity. One wonders whether such food production ought to be stimulated in areas where a part of the population suffers from malnutrition. Dramatic increases in the quantities of carnivorous fish and shrimps, produced by farming the sea, cannot be expected, not only because of the luxury character of the end-product, limiting the market, but also because one would have to compete for fish meal with pig and chicken farmers, who because their product gives a much better conversion rate than shrimps and eels, are in a stronger position. The possibility remains, however, that in due course the resources of the Antarctic krill will be tapped, as once was done by whales, or the small fish from the scattering layer will be used, in order to raise the production of fish meal from items unsuitable for human food.

Despite all these facts, i.e. the confinement of sea farming to sheltered coastal areas, the need for unpolluted water, the conflict with recreational demands and inshore construction activities, the huge quantities of formulated feed partly consisting of fish meal that one would need for a drastic increment in the production of farmed carnivorous species, one feels compelled to conclude that farming the sea is an important industry. It contributes to man's welfare and well-being although the impact of its total production on the global food situation cannot possibly be very great. Raising the contribution of food from the sea from the present 1% to a level of 1.1% by farming procedures alone, would require an extra production of about 1 000 000 tons of farmed crop, a figure which is surely not within easy reach.

BIBLIOGRAPHY

Bardach, J.E., 1968. Aquaculture. Science, 161: 1098—1106.

Bardach, J.E., Ryther, J.H. and McLarney, W.O., 1972. Aquaculture. The Farming and Husbandry of Freshwater and Marine Organisms. Wiley Interscience, New York, N.Y., 868 pp.

Baughman, J.L., 1948. An annotated bibliography of oysters. Texas A & M Research Foundation, College Station, Texas, 794 pp.

Dassow, J.A. and Steinberg, M.A., 1973. The technological basis for development of aquaculture to produce low cost food fish. Mar. Fish. Rev., 35 (11): 6—13.

Glude, J.B., 1966. Criteria of success and failure in the management of shellfisheries. Trans. Am. Fish. Soc., 95: 260—263.

Harada, T., 1970. The present status of fish cultivation research in Japan. Helgol. Wiss. Meeresunters., 20: 594—601.

Hempel, G., 1970. Fish farming, including farming of other organisms of economic importance. Helgol. Wiss. Meeresunters., 21: 455—465.

Hidu, H., 1971. The world culture of marine mollusks. Annu. Rep. 1970, Am. Malacol. Union, 18—25.

Imai, T., 1967. Mass production of molluscs by means of rearing the larvae in tanks. "Venus". Jap. J. Malacol., 25: 159—167.

Iversen, E.S., 1968. Farming the Edge of the Sea. Fishing News (Books), London, 301 pp.

Joyce, E.A., 1972. A partial bibliography of oysters with annotations. Spec. Sci. Rep. No. 34, Mar. Res. Lab. Fla. Dep. Nat. Res., St. Petersburg, Fla., 846 pp.

Korringa, P., 1969. Shellfish of the North Sea. Serial Atlas of the Marine Environment, Folio 17, Am. Geogr. Soc., New York, N.Y.

Korringa, P., 1970. The basic principles of shellfish farming on the continental coast of Europe. Symp. Mollusca, Mar. Biol. Ass. India, 1968, Part III: 818—823.

Korringa, P., 1971. Mariculture. McGraw Hill Yearbook of Science and Technology, McGraw Hill, New York, N.Y., 1971, 12—23.

Loosanoff, V.L., 1965. Mariculture, its recent development and its future. Agric. Eng., 46 (2): 73, 93—97.
Loosanoff, V.L. and Davis, H.C., 1963. Rearing of bivalve mollusks. Adv. Mar. Biol., 1: 1—136.
Mason, J., 1971. Mussel cultivation. Underwater J., April 1971: 52—59.
McNeill, W.J., 1970. Marine Aquaculture. Oregon State Univ. Press, Corvallis, Oreg., 172 pp.
Milne, P.H., 1972. Fish and Shellfish Farming in Coastal Waters. Fishing News (Books), London, 208 pp.
Parssons, T. and Takahashi, M., 1973. Biological Oceanographic Processes. Pergamon Press, Oxford, 186 pp.
Ranson, G., 1952. Les huîtres, biologie-culture; Bibliographie. Bibl. Inst. Océanogr. Monaco No. 1001, 134 pp.
Webber, H.H., 1968. Mariculture. BioScience, 18: 940—945.
Yonge, C.M., 1960. Oysters. New Naturalist Special Volumes, No. 18, Collins, London, 209 pp.

Chapter 1

"NORI" FARMING IN JAPAN (CULTIVATION OF EDIBLE SEAWEEDS OF THE GENUS *PORPHYRA*)

Example

First stage: Yamaguchi Prefectorial Marine Products Seedling Centre at Aeo-cho, Yoskiki-Gun, Yamaguchi Prefecture. Second stage: Miyoshi Brothers at Nijima Aeo, Yamaguchi ken. Period of observation: 1972—1973.

A. BACKGROUND

A.1. General principles

The first link in the food chain, the organisms forming the primary production, may constitute about 85% of the food man gathers on the land, but of what he takes out of the sea it forms only a minor fraction. Since olden days certain species of edible red-weeds have predominantly been collected, especially in Japan and China, and to some extent also in Great Britain and Canada. About 300 years ago the red-weed *Porphyra*, locally called "nori", began to be farmed in Tokyo Bay. The observation that red-weed settled on the brushwood of fish traps led in due course to large-scale planting of brushwood especially for this purpose. The rulers of the Tokyo Bay area had the monopoly for nori farming for a long time, and only from 1868 onwards, when the feudal regime had ended, was there freedom to practice nori farming elsewhere along the Japanese coastline. Bundles of brushwood, oak and leafless bamboo were dug into the sea bed in holes prepared with a special tool, a conical wooden frame with two long upright handles, pushed down by the farmer (Fig. 1). Experience showed that the onset of autumn was the best time to plant the brushwood, and that harvesting the *Porphyra* fronds should be concentrated in the winter months. Careful observation taught the farmers that the settling of nori was best in water of high salinity, whereas growth was faster in water of reduced salinity, nutritionally enriched to some extent by run-off from the arable land. Transplantation of branches carrying the young specimens of *Porphyra* therefore led to an improvement in farming results.

It was in the late 1920's that another type of collector, large-meshed nets of coco-fibre, strung out horizontally between rows of bamboo poles, was introduced. Though coco-fibre was later replaced by other materials, predominantly synthetic nowadays, net-farming soon became the basis of nori production in Japan.

Fig. 1. Planting bundles of brushwood as collector for Porphyra. (After Smith, 1905.)

Since the adult *Porphyra* fronds die off in early summer, the origin of the spores from which the new generation settled on nets and brushwood placed in autumn was a mystery. Practical experience taught the nori farmers that in certain sites there was better and more consistent settling than in others, and above all that it is of paramount importance to place the nets at a well-defined level relative to the tides. Step by step development of the methods improved the results, for example varying the level of the nets in the course of the development of the newly settled red-weeds. For the first month the best level appeared to be about 84 cm above the low water spring tide level. This level promises the most dense settling. Some time after this the net is placed 20 cm higher for a few dry and warm days in order to kill weed organisms among which are green algae of the genus *Enteromorpha.* Next a level about 35 cm lower is selected, which promises excellent growth, and about 10 days later the nets are raised again 25 cm to obtain strong dark-red fronds, ready for harvesting after a few days.

Gradually the production increased, but not without ups and downs due to occasional poor settling and the devestation of disease and floods.

A real break-through came in the year 1949. It was then that a British algologist, Miss Kathleen M. Drew, published her findings on the biology of the British species *Porphyra umbilicalis.* She demonstrated that the carpospores produced by the fronds develop during the summer season as perforating algae in molluscan shells and that the monospores produced in autumn by this perforating phase — called the Conchocelis phase — are the spores which settle on rocks or brushwood and develop into new *Porphyra* fronds. This knowledge of the life-cycle soon turned out to be of great practical importance for the Japanese nori industry. It was the Japanese biologist Kurogi who, prompted by Miss Drew's findings, started investigations on Japanese species of *Porphyra*. He soon came to the conclusion that these have a similar type of life-cycle, and he managed to rear the various phases under controlled conditions in his laboratory. From that moment onwards the importance of molluscan shells as a substratum for the Conchocelis phase of nori was realised. Soon the farmers learned to seed oyster shells with the carpospores produced in early summer, to watch and control the growth of the Conchocelis phase which lives just under the surface of the white innerside of oyster shells, to place such shells in large numbers in the sea at the time the monospores are about to appear, and to place several layers of netting over them in order to achieve dense settling on the net material.

In a later phase it was found to be more efficient, with more consistent results, if settling of the monospores on the nets was allowed to take place in tanks, and nowadays this is the system widely adopted in nori farming. Some farmers still prefer to grow both stages of *Porphyra* themselves, bringing the nets with newly settled monospores to racks in the sea, but the vast majority of the nori farmers have learned that rearing the Conchocelis phase is best left to the Prefectorial Seedling Centres. They can bring their nets to the Centres in the appropriate season to receive a guaranteed good settling of monospores, and then take these nets to the sea where the edible fronds of *Porphyra* will grow in a short time.

A further improvement followed from the discovery that nets carrying newly settled monospores can be stored at low temperature for quite some time. Refrigeration and even deep-freezing prevents growth, but keeps the young algae alive. Thus one can prolong the harvesting season by growing two crops one after the other in the same place. Nori farming is now a profitable business since the price paid for dried nori is high. A shortage of sheltered sites in which to build nori racks gradually developed, mainly because several estuaries became too polluted for nori farming. The next step was the use of moored rafts, constructed in such a way that the nets carrying the red-weeds emerge at low tide for the optimal length of time, just as on the racks.

The fronds harvested by the nori farmers are not delivered to the dealers fresh, but are invariably processed by the farmers themselves into dry sheets of standard dimensions. Formerly this was all done by hand, but now it is done predominantly by specially designed machines.

Nori farming in Japan is an industry of considerable economic importance. Thanks to various farming improvements introduced in recent years, the

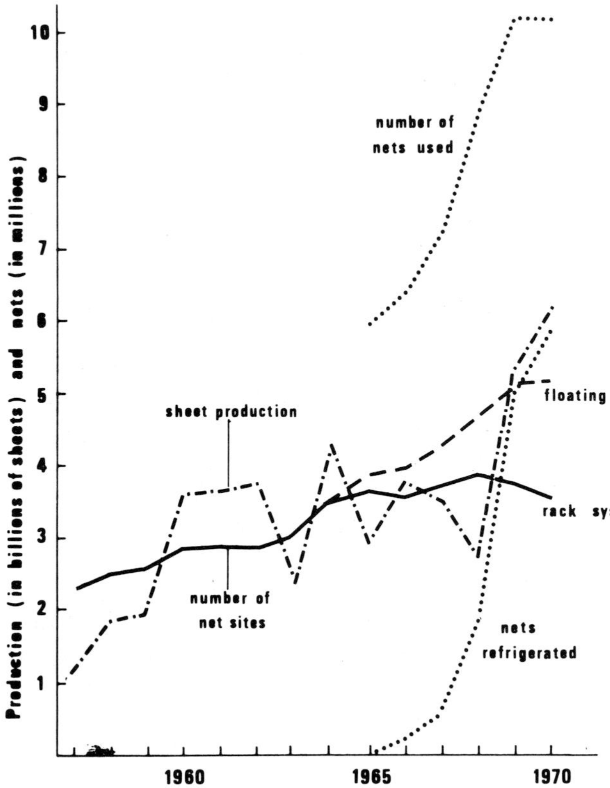

Fig. 2. Trends in nori farming. (After S. Suto, 1971.)

production of sheet rose steeply, from about 10^9 sheets in 1956 to 6×10^9 sheets in 1970 (Fig. 2). Since one standard sheet, weighing 2.5 g, is sold for a price of about Y 12, the total value of the nori crop is impressive. Few will realise that nori ranks highest when one lists the various crops harvested from the sea according to their monetary value.

The number of nori farmers in Japan is now about 60 000, many of them working at their nori farms only in the winter season. The total number of nets of standard size now used is about 10^7 per year, for which one requires about 60 000 ha of shallow ground. The main areas for nori farming are the northern shore of the Inland Sea (Seto Naikai) west of Osaka, the north and west coast of Kyushu, also Ise Bay, Tokyo Bay, the northeastern coast of Honshu, and Uchiura Bay on Hokkaido. Tokyo Bay has lost much of its former importance as a nori farming area due to pollution.

Though the Japanese use a great number of algal species as human food, *Porphyra* (nori) is by far the most important, representing 75% of the total value of all algae harvested for human food and industrial purposes in Japan.

Recently the farming of two species of brown-weed suitable for human consumption, viz. *Laminaria japonica* ("konbu") and *Undaria pinnatifida* ("wakame") has started, but these efforts have hardly passed the experimental phase.

A.2. Biology of the species

The red-weed *Porphyra* has a very thin purplish thallus, elongated in shape, often with lobed fringes. 20 species from Japanese waters have been described, but by far the most important for farming are the true amanori (Fig. 3), also called asakusa nori, *P. tenera* Kjellman and the susabi nori, *P. yezoensis* Ueda. The former species yields a product of the highest quality, but is somewhat susceptible to disease, so that the farmers often prefer to grow the latter species, which is much hardier. It is now the dominant species in the nori industry. Hybrids can be farmed, and sometimes also *P. angusta*, because it can be harvested very early in the season, but this species is rather susceptible too, whereas the dark-coloured fast-growing *P. pseudolinearis* has a poorer flavour.

The thallus of *Porphyra* develops after the monospores have settled on some type of substratum. Settling can be expected in September or October when the days shorten and when the water temperature has dropped well below 23°C. In the first few days after the full or new moon the greatest settling intensity invariably occurs. *Porphyra* evidently belongs to those

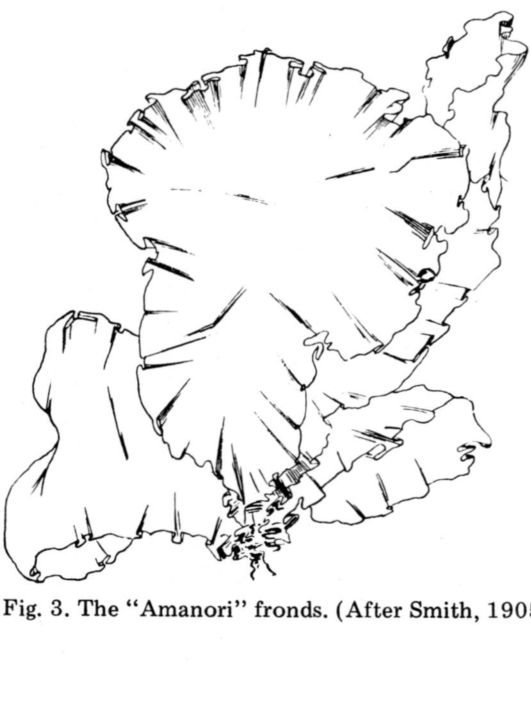

Fig. 3. The "Amanori" fronds. (After Smith, 1905.)

organisms which show a lunar periodicity in reproduction with peaks of activity at both spring tides. The monospores can settle on brushwood even when this is already coated to some extent with a film of diatoms. *Porphyra* fronds grow remarkably rapidly; once the thalli have reached a size of 1 cm they can be expected to reach a length of 30 cm in 2 to 3 weeks. Specimens often grow up to 45 cm long. In the winter season when water temperatures are about 10° C, this short-day plant grows about 10 to 15 cm in one tidal cycle (15 days). The farmers usually avoid the risk of losing the fronds in rough weather; this loss may occur as a consequence of keeping the fronds too long on the nets in an effort to grow them to maximal length. When cut off, the remaining parts of the fronds easily grow again, so that one can harvest three or four times from the same initial material.

When the water temperature rises in spring, and the days lengthen, the fronds produce reproductive organs along their fringes. Some plants produce carpogonia, others spermatia. The fusion of these yields the carpospores which are somewhat amoeboid and require the calcareous shell of a mollusc for their further development. They germinate and then penetrate into the innermost layer of the calcite-ostracum of the shell. Here they sprout prolifically, out of reach of all kinds of browsing organisms, but sufficiently near the surface to receive sunlight for photosynthetic activity. Individuals in this perforating phase gradually become so dense that their presence can be detected by the naked eye as greyish-purplish spots, in due course about 1 cm diameter. Such plants, recognisable as dark red incrustations, were described by Batters in 1892 as the perforating alga *Conchocelis rosa*, long before Miss Drew made clear that they are part of the life-cycle of the red-weed *Porphyra*. The Conchocelis phase is a long-day plant which grows best at water temperatures over 25°C and at a light intensity not higher than 500 lux, but otherwise is not very exacting. It does however take advantage of applications of nutrients such as potassium phosphate and potassium nitrate. When the days shorten and the water temperature drops below 23°C the Conchocelis phase begins to produce non-motile monospores, breaking through the shell again to release them in the water. For a few successive days each Conchocelis plant produces hundreds of spores, especially around sunrise. The tides help to carry the monospores to a good substratum on which they settle and produce in due course a frond. Under controlled conditions, one can manipulate light intensity and water temperature in order to obtain sporulation when one wishes, earlier if so desired than sporulation in the wild. This opens the possibility of a longer harvesting season.

A.3. Geographical situation

The nori farm of the Miyoshi Brothers is to be found in Yamaguchi Prefecture, in the southwestern tip of Japan's main island Honshu, not far from the city Yamaguchi. The headquarters of the enterprise, where the nori is

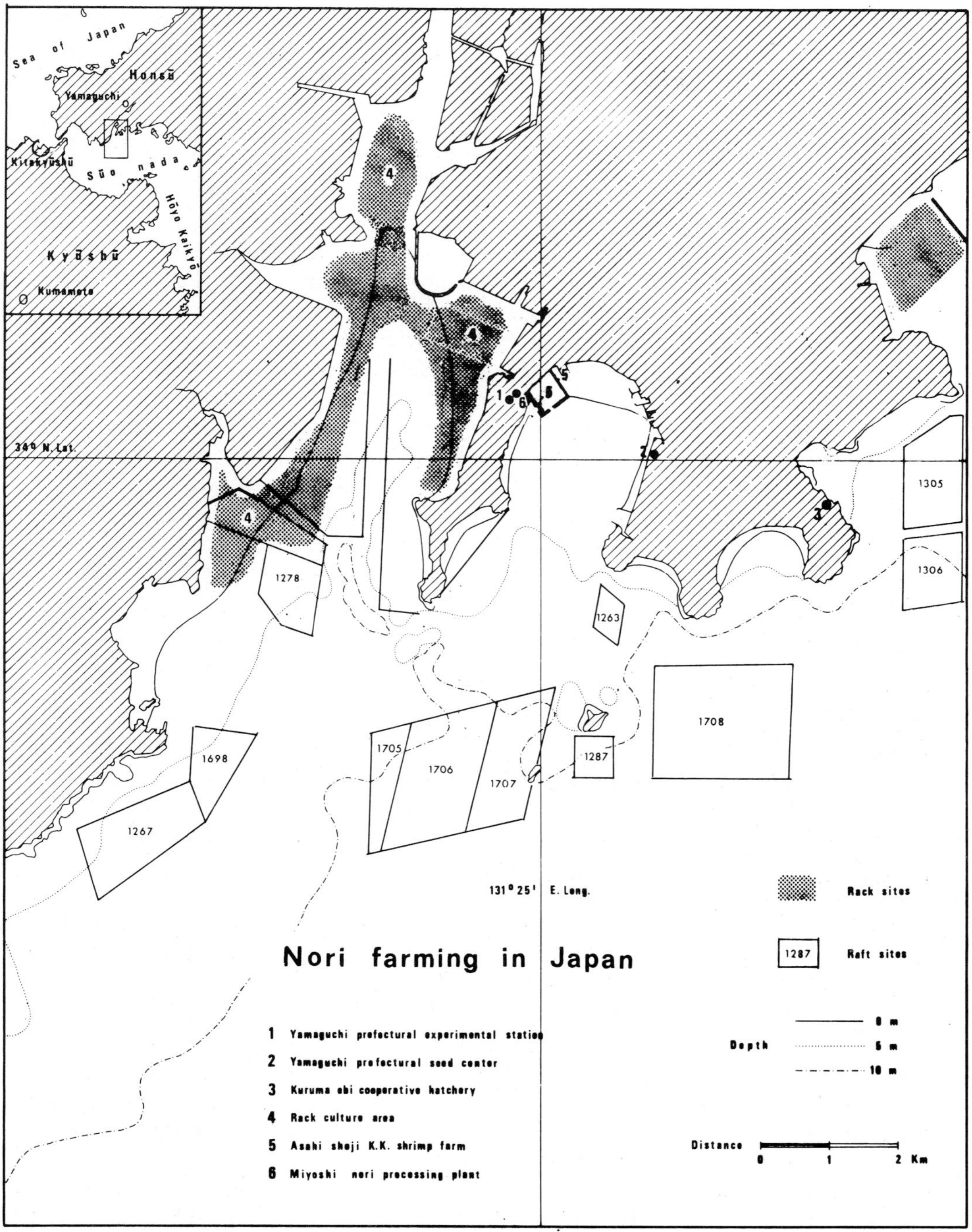
Sea of Japan
Honsū
Yamaguchi
Kitakyūshū
Sūo nada
Hōyo Kaikyō
Kyūshū
Kumamoto
34° N. Lat.
131° 25' E. Long.
1278
1263
1305
1306
1708
1287
1705
1706
1707
1698
1267
Rack sites
Raft sites
Nori farming in Japan
1 Yamaguchi prefectural experimental station
2 Yamaguchi prefectural seed center
3 Kuruma ebi cooperative hatchery
4 Rack culture area
5 Asahi shoji K.K. shrimp farm
6 Miyoshi nori processing plant
Depth
0 m
5 m
10 m
Distance
0
1
2 Km

processed into sheets, is on a narrow peninsula due south of the town of Ogori at the exact geographical position 34° 00′25″ N, 131°24′40″ E. It is on the northern shore of the western prolongation of the Seto Inland Sea, called Suo Nada (see Chart).

A.4. Hydrographic pattern

For the old style of nori farming, based on the settling of monospores originating in the open sea, one had to select sites of high salinity for placing the virginal nets or the brushwood, in order to obtain a rich settlement. When one works with seeded nets one may place these in water of slightly reduced salinity, that is, water usually with an admixture of run-off water from arable land, quite rich in nutrients, and often somewhat polluted. Under truly estuarine conditions care must be taken not to expose the nets for too long a time in the upper layers of the water when the salinity of these is too low.

Tides and waves. Since *Porphyra* is an intertidal species it is of considerable importance to select the optimum level for stringing out the nets. For capturing the monospores in the wild the optimal level is 84 cm above low water, but for growth of the fronds on the seeded nets one should place them about 70 cm above the spring tide low water level. At deeper levels growth may be good but diseases may strike more easily than at higher levels. If *Porphyra* is exposed at low water for a long time, the fronds may become dark red, and also slightly tough.

The tides are of the semi-diurnal type, albeit the tidal range (2.5 m at spring tides, 0.8 m at neap tides) during the farming period is greater for the night tide than it is for the day tide. The farmers aim for an average of 2 h regular exposure of the nori nets at low tide. At spring tide this may mean 2 h during the day tide and up to 4 h during the night tide. To obtain optimal results in nori farming it is worthwhile to study the local pattern of the tides in considerable detail, and to lower or raise the strings on the bamboo poles supporting the nets, according to the needs of the nori. The "Nori-Times", appearing at regular intervals, gives the farmers information on the tides expected in the weeks to come. The tidal currents are not very strong, but waves whipped up by the wind may become dangerously high in places with insufficient shelter. Fortunately the coast is well-indented so that the farmers can easily select sites for rack building with sufficient shelter against the prevailing winds.

The racks are built close inshore, not only for better shelter but also because it is difficult to build racks where the depth exceeds 5 m. A special case is raft cultivation, developed on account of an increasing shortage of suitable shallow grounds for racks. The nori rafts are moored in water up to a depth of 25 m. Nevertheless, the nori, at least during the first 3 weeks of cultivation in the sea, are exposed to the desired intertidal conditions. Instrumental in this are huge plastic frames which can carry up to 20 nets and which can be handled in such a way that the nets are exposed to air for at least 2 h on alternate days, to simulate the tide.

Water temperature. Nori farming is seasonal in character and escapes the high water temperatures of summer. The water temperature should not exceed 20°C during the growth of the thalli. Temperatures from 10 to 15°C support the fastest growth, but temperatures as low as 3°C can easily be endured by *Porphyra.*

Salinity and nutrients. The growing fronds of *Porphyra* can withstand reduced salinities, provided that prolonged contact with water of 10% salinity or lower is avoided. That one observes a faster growth at reduced salinities should not be ascribed to the salinity as such, but rather to a higher content of nutrients, brought from arable land. Some rain is therefore an advantage to the nori farmers. But over-low salinities constitute a dangerous situation, because they will reduce the resistance of the red-weeds against disease. One could consider fertilizing the water around the nets with inorganic fertilizer, but the tides would wash too much fertilizer away into areas not exploited by the farmers who went to the trouble of spreading it. It is preferred therefore to build the racks or to moor the floats in places where there is slight pollution with domestic sewage. Under such conditions growth is excellent. But industrial pollution is shunned, since it may seriously endanger the health of the red-weed or render the nori unfit for human consumption.

A.5. Legal aspects

The nori farm of the Miyoshi Brothers is a private enterprise, but there is close cooperation with the Prefectorial Seedling Centre at Aeo-cho where they bring their nori nets to be seeded in the appropriate season. According to the Japanese law, the State owns the marine environment up to the high water line, but one can lease sections of sea bed for farming purposes. Since the total area of suitable sites for nori farming is not unlimited, the fishery authorities do their utmost to arrange a fair distribution of the suitable grounds among the interested farmers, ensuring that everyone is given some ground covering a range of qualities and up to a given maximum area. There is no rental charge, but the local cooperative collects 3% of the annual yield from each nori grower as a contribution.

In the year of observation, the Miyoshi Brothers had no rack sites allotted to them but only raft plots. In the future they may be given tidal plots to use for racks, as has happened in previous years.

The Miyoshi Brothers had, at the time of observation, 120 nets at sea with a total net area of about 3 200 m^2. The site area one needs is roughly 13 times the net area. Thus the Miyoshi Brothers' enterprise occupies about 4 ha. They have 4 plots of 30 nets each in the following raft areas: 1707, 1708, 1263 and 1287, which are indicated on the chart.

Since the nori crop is valuable it has to be guarded against theft and poaching. Surveillance is carried out on a mutual basis. Whenever members of the cooperative go out to a plot they will as a rule check the other plots as well.

B. SEQUENCE OF OPERATIONS

The first phase of nori farming, which ends with the seeding of the nets is carried out locally by the Yamaguchi Prefectorial Marine Products Seedling Centre at Aeo-cho. After the nets have been seeded, it is the responsibility (in our example) of the Miyoshi Brothers at Nijima Aeo to continue the farming operations, which ultimately lead to the sale of dried sheets of amanori.

B.1. Collecting the carpospores and seeding the shells

At the end of the growing season, from late February to early April, the nori leaves on the nets at sea are inspected for developing sporangia, which are found on the fringes of the fronds. If sporangia are present, ripe leaves of the highest quality are carefully selected from the regular commercial crop. On arrival at the Yamaguchi Seedling Centre the ripe leaves are put into a 20-l open container filled with sea water, and the fronds are lightly squeezed to stimulate the release of the spores. After about 10 min one can be sure of a high concentration of spores in the supernatant water.

These carpospores should then be brought into contact with shells. For this purpose, clean oyster shells are needed. They are carefully washed in fresh water and placed with the white inner surface upwards on the bottom of a shallow concrete tank. One can then use a watering-can to pour the liquid containing the carpospores evenly over the shells. About 1 kg of ripe nori leaves yields sufficient spores to seed about 20 000 shells. Within 2 days the somewhat amoeboid carpospores will germinate and penetrate into the shells. Good results are obtained at water temperatures from 10 to 15°C. Samples of the shells are inspected under the microscope; if 50 to 100 spores have perforated each of the shells, they are ready for the next phase.

At some farms the ripe nori leaves are simply laid on top of the shells, covering from 10 to 20% of them, and left some time for penetration. The leaves must be taken away before they begin to decay.

B.2. Preparing the seeded shells for rearing the Conchocelis phase

The best development of the Conchocelis phase can be expected when the seeded shells are not in fact placed on the bottom of a tank, but are hung out freely in sea water. To do this holes are drilled in the shells, and they are strung on nylon cords after seeding. At the Yamaguchi Seedling Centre two

Fig. 4. Strings of oyster shells as collector for the Conchocelis phase. (Photo P. Businger.)

holes are usually drilled in every shell to make strings of 15 shells, with the tips touching (Fig. 4). Elsewhere knots are tied in the cords to keep the shells, in which a single central hole has been made, separate from one another.

B.3. The rearing tanks for the Conchocelis phase

The Yamaguchi Seedling Centre has a special building for rearing the Conchocelis phase. It consists of one large room with 24 identical concrete basins of 2 × 3 × 0.70 m, placed in two rows. The walls and roofs are provided with windows, but black curtains are installed at every window to allow control of light intensity. There are pipelines with valves for filling the tanks with sea water and for renewing the water when necessary. 16 bamboo sticks are placed lengthwise over each tank, just above the water level, and about 35 strings of shells are hung loopwise over the sticks. Thus the shells are evenly distributed in the tanks, each tank holding about 8 000 shells, and each shell bathed in sea water freely on all sides. In all, the building can accomodate about 200 000 shells for Conchocelis rearing.

This phase, the preparation and seeding of the shells, making the strings, and hanging them out in the basins, is quite laborious. In the Yamaguchi Seedling Centre it keeps five men in full-time employment for about 2 months.

B.4. Rearing the Conchocelis phase

Rearing the Conchocelis does not require much labour. The perforating algae are not exacting in their requirements, and there is no need to aerate the tanks. It is however important to control both water temperature and light intensity carefully. The temperature should rise in early summer and

Fig. 5. Rearing tanks for the Conchocelis phase. Note bamboo sticks in the foreground and curtains to adjust the light intensity. (Photo P. Businger.)

should not drop below 23°C too soon as this provokes premature spore formation. The light intensity should be maintained at 500 lux or slightly less. In the autumn both temperature and illumination can be manipulated in such a way that the spores mature at the needed time. As a rule there is no need to renew the water in the course of the summer season, but if the quality of the water deteriorates then it is done. Usually the shells and the sea water in the tanks contain enough nutrients to feed the Conchocelis plants, but careful administration of a small dose of soluble phosphate and nitrogen compounds may stimulate the development of the algae.

The shells remain for about 5 months in the indoor tanks (Fig. 5) and in due course the Conchocelis plants are seen as circular spots, greyish purple in colour, with the naked eye. In shells found in the wild the Conchocelis plants are not so easily seen because such shells are host to a variety of perforating organisms, among which *Gomontia* may dominate.

There are still nori farmers who prefer to rear their own Conchocelis but their results are less consistent than those of the better-equipped Seedling Centre. Sometimes the shells become overgrown with diatoms, which may hamper the development of the Conchocelis plants. In that event the shells can be brushed clean, since the Conchocelis lives safely inside the shell material.

B.5. Production of monospores

In September or October the Conchocelis plants are ready to produce monospores, which should of course leave the shells. The sporangia break through the shell and release the monospores freely into the sea water. A drop in temperature well below 22°C and a shortening of the day-length provoke maturation, and the experienced nori farmer knows how to govern this process by manipulating temperature and light.

B.6. Seeding the nets with monospores

The non-motile monospores should be carried by the currents to some suitable substratum. From the very beginning, nori farmers have understood that the nori production could be increased greatly by the timely offering of such a substratum. By tradition they call the substratum "hibi", irrespective of whether it is bundles of brushwood, split bamboo blinds, or nets.

At the Yamaguchi Seedling Centre nets of synthetic fibres 2 to 3 mm in diameter are used exclusively. The nets of standard size are here 18.20 m long by 1.50 m wide, with meshes of about 15 cm square (30 cm stretched).

For seeding the nets a shallow concrete outdoor tank measuring about 7 × 6 × 0.50 m containing six rotating drums made of round iron bars is used (Fig. 6). The drums, placed vertically, are 1.50 m in diameter and measure 2.00 m across. They are driven by an electric motor with hydraulic power transmission, and turn slowly at about 2 r.p.m. The undersides of the

Fig. 6. Rotating drums for seeding nori nets. (Photo P. Businger.)

drums are about 25 cm above the bottom of the tank.

To operate the system one winds 30 nets of standard size around each drum. The oyster shells carrying the mature spores are strewn on the bottom of the tanks without being removed from the nylon cords. When the spores are released from the sporangia the water in the tank acquires a greenish brown colour. Then the drums are rotated, keeping the water in which the monospores are suspended sufficiently turbulent to prevent the spores from settling out on the bottom of the tank. When a spore touches the netting it adheres to it in a very short time. When everything has been properly set up it takes from 20 to 60 min to seed the nets this way. Inspection under the microscope confirms whether or not the operation has been a success; using a magnification of 100 × about 50 spores attached to the netting should be seen in one field of vision. If this concentration is found, the nets are ready to be taken to the sea. The whole operation of seeding 30 nets, using one drum, takes 2 men about 3 h. The Yamaguchi Seedling Centre seeds about 20 000 nets in the height of the season. Since they can have up to 200 000 shells in their Conchocelis tanks, one can safely conclude that ten shells produce more than enough spores to seed one net.

B.7. Other systems (Fig. 7)

The description above refers to the system used by the Yamaguchi Seedling Centre, but other methods may be used:

(a) Stretching the virginal nets (or widely interspaced split bamboo blinds) horizontally in the sea between poles, and hanging out here and there between them perforated sacks filled with shells bearing Conchocelis.

(b) Spreading shells bearing Conchocelis plants in the intertidal zone of

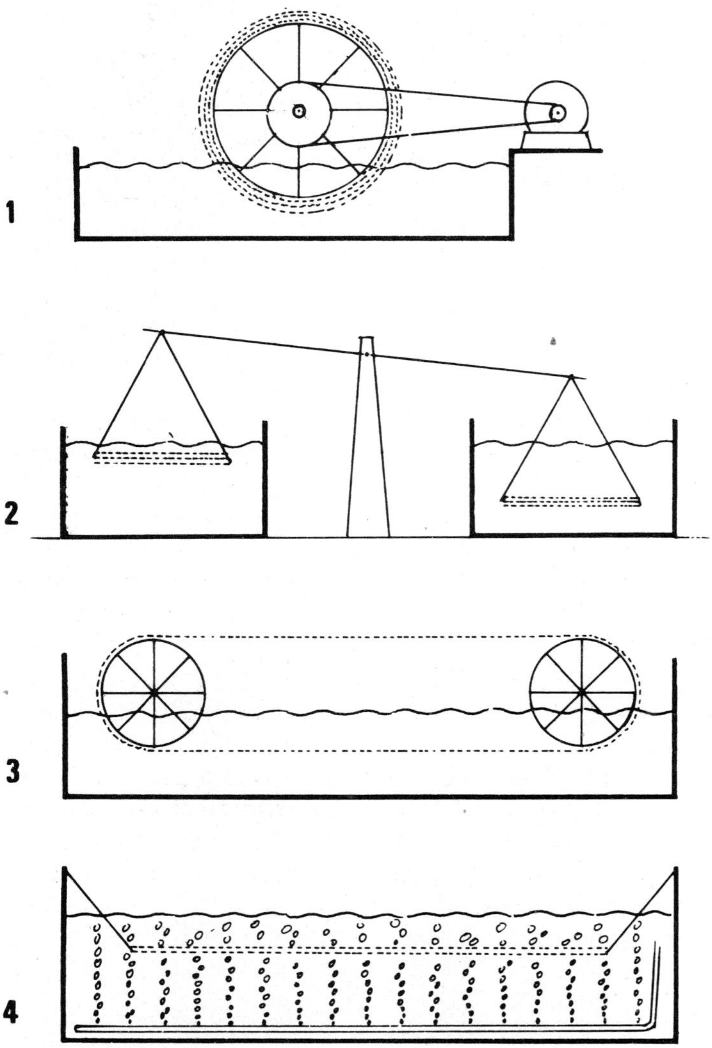

Fig. 7. Different techniques for seeding nori nets: (1) Rotating drum; (2) scales moving up and down; (3) double drums; (4) air-bubble system.

the sea bed and putting five or six layers of nets on top of them. After a couple of days these nets are brought to the racks for further growth.

(c) Strewing the shells in a concrete tank, the bottom of which is overlain with a great number of thin pipes through which air can be blown. The nets are suspended horizontally above the bottom, and the air bubbles cause the spores to rise towards them.

(d) Folding up the nets and placing them on the pans of a large balance, which is moved slowly up and down by an electric device. The nets remain permanently under water and the water in which the spores are suspended is kept sufficiently stirred.

B.8. Storing the seeded nets

In the Yamaguchi Seedling Centre there is a large concrete basin which is adjacent to the seeding tanks and is subdivided into sections in which the newly-seeded nets can be stored safely until the farmers come to pick them up. Formerly, the seeded nets were taken to the sea as quickly as possible, but nowadays it may be preferable to store some of the seeded nets for a time. This allows the farmers to extend the harvesting season, by using two sets of nets successively on one rack, and moreover, to keep seedlings in reserve in case some misfortune should strike the nets already set out in the sea.

The newly seeded nets can safely be placed in cold storage, provided they are folded up and covered with polyethylene sheets to keep them moist. In

Fig. 8. Seeded nori nets stretched horizontally between bamboo poles. (Photo courtesy Yamaguchi Prefectorial Marine Products Experimental Station, Aeo.)

this way one can store the nets for several weeks. Young nori buds will also survive if the nets are first of all dried down to 20 to 30% moisture, and then packed in polyethylene bags for storage at —20°C. By this method the seedlings can be kept alive for more than 6 months, and at the same time weed organisms such as diatoms are killed off.

B.9. Bringing the nets to the racks

The farmer should build racks to receive the seeded nets before he brings the nets to the Seedling Centre. He selects one of the sites available to him, preferably well-sheltered and with current velocities of 30 cm/s or more at the height of the tide, and builds the racks, which consist of two rows, each of nine long bamboo poles, driven into the sea bed with a powered pump and a rubber hose, forcing water under pressure through a nozzle. The poles are placed in such a way that a stretched net fits precisely between the two rows. Usually one places four or five of such "sets" next to each other so that most of the rows of bamboo poles serve two nets. For five nets one needs thus only 54 poles (Fig. 8).

Fig. 9. Floating frames of plastic tubing to be installed on a raft after receiving a set of 20 seeded nori nets. Such frames serve to imitate the tide on a raft. (Photo P. Businger.)

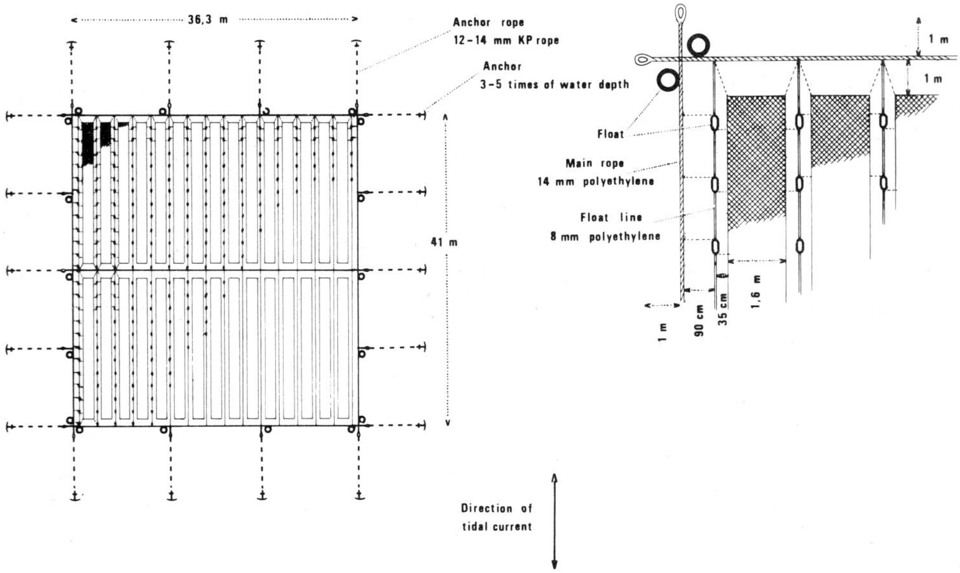

Fig. 10. Nori raft of nylon ropes.

The Miyoshi Brothers put out 100 to 150 nets per year. They can take the folded-up seeded nets on the deck of their 5 m long wooden motor boat, and then string them out between the poles. This is a job which is best done with a team of three workers, making use of small rectangular plastic boats taken along with the motor boat.

When one has to operate on raft sites because of a lack of suitable intertidal grounds for rack building, it has to be taken into consideration that the nori require to be exposed to air from time to time, especially during the first 3 weeks of their sojourn in the sea. Floating frames, 18 m long, made of plastic tubing, can be used for this purpose (Fig. 9). Up to 20 nets can be attached to one single frame. Having been brought to the site, the frame is placed in a slot in the raft and adjusted in such a way that the nets are exposed to air for at least 2 h every other day, or whenever possible, every day. This operation, which takes at least 3 weeks in succession, requires the attention of the whole Miyoshi team working full time. When the nori has grown to a size of about 2 cm the nets can be placed one by one into the slots in the raft.

Rafts may be constructed in the form of a wide light bamboo framework. This offers the possibility of attaching the nets directly to the frame in one single layer and the tide can be imitated by placing styrofoam floats under the frame to lift the net temporarily out of the water. If the raft consists however of a frame of nylon ropes, duly anchored at all sides and supported by floats, one has to use the plastic frames to imitate the tide during the first 3 weeks at sea. Later the nets can be strung out in one single layer. Such a nylon raft has 30 slots which can each accommodate a single net (Fig. 10).

B.10. Growing the nori fronds

In the first month the nori seedlings grow slowly, but from the moment the buds have reached a length of about 1 cm one can expect much faster growth — 10 to 15 cm in a tidal cycle of 15 days. About 50 days after seeding the nets the fronds will have a length of 15—20 cm. It is during this period that the farmer should exercise the greatest care, adjusting the level of his nets according to the tide, and taking into account such factors as water temperature, salinity and disease. The farmer receives all the scientific advice he needs in this period and can easily raise or lower his nets by moving the supporting ropes up or down over the bamboo poles.

B.11. Harvesting the nori

Most of the harvesting takes place from December to March, nori being a typical winter crop. Formerly the only way to collect the fronds was to pick them one by one from the brushwood, or from the bamboo blinds or netting, a slow and cold job.

Fig. 11. Rotary cutter installed in plastic boat lined with finely-meshed netting. (Photo P. Businger.)

Modern farmers such as the Miyoshi Brothers use mechanised equipment. In one of their small rectangular plastic boats (about 1.80 m long), they have installed a rotary cutter (Fig. 11). This is an apparatus not unlike a lawn mower, but with a length of 1.50 m, covering the entire width of the net. At the front side it is shielded, and sharp blades rotate in the long cylinder under a row of 17 curved bars which prevent the net from being cut. A small gasoline engine powers the cutter. Two men in the boat pull the net over the cutter, and a third worker stays in the motor boat. At low tide the net height can easily be adjusted to facilitate this work. The cut fronds fall into a finely-meshed nylon net placed over the bottom and walls of the boat and are easily collected.

When nori is harvested from a raft the same mechanised equipment can be used. Here too, the rotating cutter is passed under the net. The harvesters untie and retie the lateral ropes of the net as it passes.

Within a fortnight the remaining fronds and frond stumps will again have grown sufficiently to allow further cropping. As a rule one harvests three or four times from the same net, although the crop is usually smaller every time. At the end of the harvesting season the nets should be carefully cleaned with fresh water, and dried.

Fig. 12. Washer for use on board the ship. (Photo P. Businger.)

Not all the nori farms use the same equipment to harvest the fronds. Another type of mowing machine which is moved systematically by hand underneath the nets and is equipped with a vacuum pump for collecting the cut-off leaves is quite popular too.

Immediately after harvesting, the nori is thoroughly washed in sea water. This is done on board the motor boat equipped with a cylindrical washer (Fig. 12). The washer consists of a finely-meshed cylindrical cage with an agitator inside it. The nori is fed into the hopper on the top, and a powered pump with a hose feeds the washer with ample sea water.

B. 12. Processing the nori

The nori farmers do not sell the newly harvested product in the fresh state, but invariably process it into dry sheets. In the old days this was a laborious job, sun-drying included (Fig. 13), but nowadays this phase is mechanised too. One therefore harvests the quantity of nori per day that one can handle with the available equipment. For the Miyoshi Brothers this means that in one day, enough nori to make 8 000 sheets, dried and packaged that same

Fig. 13. Processing nori in olden days. (After a Japanese print.)

day, will be harvested. Depending on the quantity to be harvested from one net (and this quantity varies according to the place and the season), one will have to cut the fronds from 10 to 15 nets each day. In other words, one standard net will yield enough nori to make 400 to 600 sheets, and each net will yield over the whole season from 1 000 to 3 000 sheets.

In the workshop ashore, the nori, corresponding to 35 to 100 kg fresh algal material, is passed along a production line. First it is washed again, in a similar washer, but this time with fresh water. Next it is fed into an electrically-driven mincing machine (Fig. 14), not unlike a meat grinder, in which it is shredded into 0.5 to 1 cm pieces. After this the nori is thoroughly mixed with fresh water, 4 kg of nori in 100 l of water, in a special tank. This nori—water mixture is then fed into the hopper of the "paper machine" (Fig. 15), a machine not unlike one used for making paper. The mixture is metered automatically into cedar-wood frames, about 30 cm square outside, into which fit mats of split bamboo, of standard size 20 × 18 cm, placed over a

Fig. 14. Mincing machine and mixing tank in the plant ashore. (Photo P. Businger.)

Fig. 15. "Paper machine" for making nori sheets. (Photo P. Businger.)

wire netting screen. About 600 cm^3 of the mixture is fed into each frame and the water drains away through the mats and the screen. The frames carrying the mats move slowly along the production line and finally pass over a heated surface. When the frames have completed their elliptical tour and return to the hopper, the mats with the drained nori are taken away, and clean mats are installed to receive in turn a measure of the nori—water mixture.

The mats carrying the nori sheets are then piled up and taken to a dryer. The Miyoshi Brothers have a gas-heated dryer (Fig. 16). During the 1 h drying cycle the moisture is reduced to 18%. To obtain a product of good quality the temperature should not exceed 50°C during drying. Some farmers use infrared dryers or oil-burning dryers.

B.13. Packing and shipping the nori sheets

After leaving the dryer the nori sheets are taken from the bamboo mats, put together in tens, and packed in bundles of 100. Then they leave the hands of the farmer who delivers them to a cooperative shipping point. Here they are carefully packed and sealed so that moisture will not reduce the quality, and shipped to all parts of the country. Of the total Japanese production of 6×10^9 sheets, Yamaguchi provides about 4×10^8.

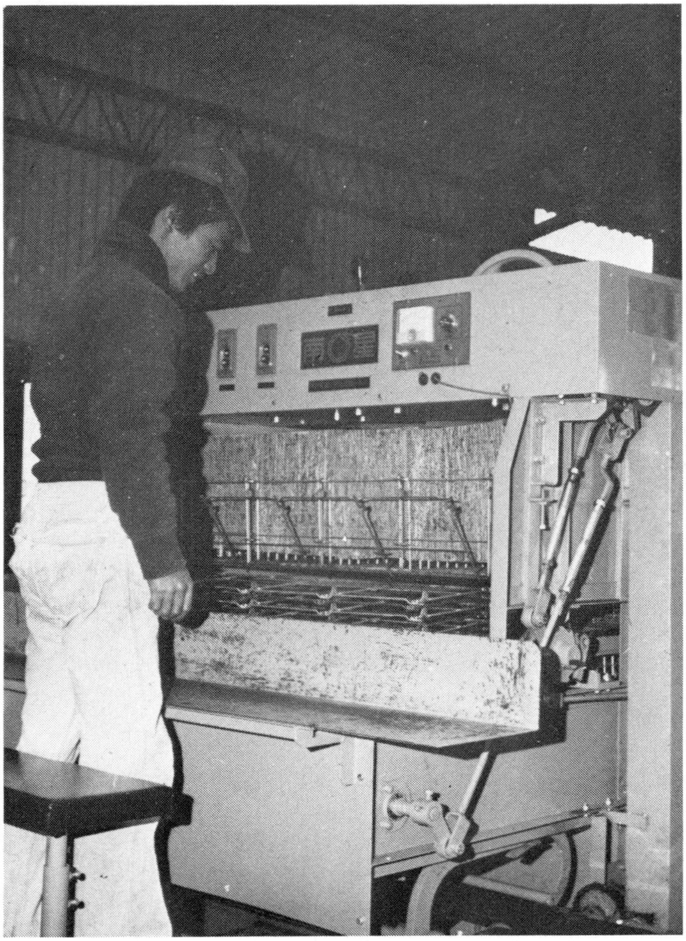

Fig. 16. Gas-heated dryer for nori sheets. (Photo P. Businger.)

B.14. Consumption of nori

One sheet of dry nori weighs 2.5 to 3 g and 25 to 30% of its weight consists of protein. Further, it contains several vitamins in notable quantities, together with a variety of mineral salts. Nori is a very popular product in Japan and fetches a high price on the market. It is however eaten in small quantities at a time, and is mainly used as a condiment, never as a principle dish. A sheet is usually toasted, to make it crisp and green. It is then broken and dipped in soy sauce or added to soups and broths to which it imparts its special flavour.

Very popular is the snack called "maki-sushi" — rice with some fish, meat or egg, rolled in a sheet of nori, and cut into slices of 3 or 4 cm thick.

C. FARMING RISKS

C.1. Hydrographic conditions

It is of considerable importance to select a good site for nori farming. Where shelter is insufficient, strong surf may cause a lot of damage, especially when the fronds are quite long. Nets strung between bamboo poles or on rafts are less vulnerable in high waves than those on bamboo blinds or brushwood.

Precipitation is a factor which can work both ways: some rain is an advantage, since it brings fresh water carrying nutrients, especially where there is run-off from rice fields. Over-reduced salinity is however to be feared, because it decreases the nori's resistance against diseases.

C.2. Predators and competitors

Strangely enough the nori farmers do not complain about predatory organisms attacking the crop, but it does happen that other weeds of a more stalwart nature settle on the nets and oust the nori. These include green algae such as *Enteromorpha* and *Ulva*, and also certain diatoms. Careful manipulation of the level of the nets can overcome this problem. Most of the competing algae cannot withstand exposure to air as much as the nori, but one should of course be careful to avoid drying out of the nori plants, as this is fatal to them.

C.3. Diseases

The nori plants are susceptible to disease, and the Japanese scientists have assessed the prevalence of ten different diseases in the Yamaguchi area. Most of these diseases are bacterial in nature, but at high water temperatures a fungus disease (*Pythium* sp.) may strike. Sometimes a noticeable part of the crop is lost, despite all the efforts of the farmer, but the Yamaguchi district does not suffer greatly from such reverses of fortune. If disease does break out, the government nori scientists advise the farmers what to do on the basis of analyses of samples. The usual remedy is to select a level for the nets which optimises the resistance of the nori, yet does not favour growth of the organisms causing the disease. Since growth is faster at greater depth, the farmers are inclined to lower the nets, but they know that disease will then strike more readily. The knowledge acquired on the disease resistance of various species of *Porphyra* is duly applied in selecting the variety to be grown. Crossbreeding experiments may improve the situation still further. If disease has been observed in a crop the farmer should exercise the greatest caution in washing and drying his nets before he uses them again. Desinfectants have not been used so far.

D. ECONOMIC ASPECTS

The first phase, growing the Conchocelis, is in the example described here entirely in the hands of the Yamaguchi Prefectorial Marine Products Seedling Centre at Aeo-cho. It is difficult to assess the economics of this government enterprise. It requires buildings with rearing tanks and laboratory facilities, outdoor tanks for seeding operations, net storage and cold storage facilities, etc. The basic material, oyster shells, is not expensive at Y 1 per shell, and only about 200 000 shells are used annually. At the peak of the spring and autumn seasons one needs many workers for preparing the shells in spring, and for seeding the nets in autumn. The Seedling Centre has set a price for seeding a net of standard dimensions at Y 250, and it treats some 20 000 nets per year. The gross income seems hardly sufficient to pay the wages, certainly not for amortisation of buildings and equipment. For the second phase, growing the nori fronds, the farmers, in our case the Miyoshi Brothers, have to face the cost of some inventory and expendable items, as follows.

D.1. Rents

The local cooperative collects 3% of the annual yield from each nori grower as payment for the use of farming plots.

D.2. Inventory items

For work in the field a motor boat built of wood, about 5 m in length, is needed. It is used for transporting workers, nets, poles and harvested nori. Such a boat, complete with engine, costs about Y 10^6 and has an approximate life of 10 years. Further, one needs rectangular plastic boats, about 1.80 m long, for attaching the nets to the poles and for harvesting the crop with the rotating cutter. Such boats cost about Y 75 000 and will last for about 10 years. A rotary cutter can be bought for about Y 50 000 and will last about 5 years. The company operates three floating plastic frames which are used on the rafts for simulating the tide during the first phase of farming at sea. Such frames cost about Y 35 000 each and will have a life of 3 years. If one works not only with racks but also with rafts one should take into account the purchase of materials for building and anchoring the rafts. A raft for 30 nets costs approximately Y 150 000 and will last about 3 years; the company operates four rafts in those years in which no rack sites have been allotted.

Ashore, the Miyoshi Brothers have a spacious shed, made of corrugated iron, in which they have installed their machines: a washing machine, costing about Y 200 000 and lasting 5 years, the mincing machine and mixing tank, also costing about Y 200 000 and lasting about 5 years, the paper maker, costing Y 300 000 with a life of more than 10 years, and the kiln to dry the sheets, costing Y 1.3×10^6 and lasting more than 10 years.

For placing poles a pump costing about Y 30 000, with an estimated life of 5 years is used; this is operated by the same 5 h.p. engine which is used to power the rotary cutter during the harvesting of the nori. In addition, some means of transporting the raw material and the marketable product, as well as transport for taking the nets to and from the Seedling Centre, is needed.

D.3. Expendable items

First of all rent must be paid for the plots leased from the government. A recurring item on the budget is the purchase of nets. A net of standard dimensions costs Y 2 000 and lasts for two seasons. The Miyoshi Brothers use from 100 to 150 nets each year. For seeding the nets every year they have to pay Y 250 per net to the Seedling Centre. Bamboo poles to support the nets last only 1 year. They cost Y 100 each and about 1 000 poles are needed for the number of nets used by the Miyoshi Brothers, provided that several sets are made alongside each other, and cold storage is used, allowing two nets in succession to be accommodated on each rack in the course of the season. For the engines of the boat, the washers on board and on land, for operating the rotary cutter, for transporting the material either to and from the Seedling Centre, from the port to the workshop or from the workshop to the cooperative shipping point, petrol is required. Various machines in the workshop run on electricity, and the kiln uses gas. If cold storage or deep-freeze facilities are to be used, these also have to be paid for.

D.4. Personnel and wages

The enterprise under consideration is owned by two brothers, who participate in the work. In fact the work is carried out by a family team since the wives of the brothers and also their father participate in operations. During the harvesting season they hire two part-time female workers to assist in the processing plant. The members of the family are not paid a regular salary, whereas the part-time workers earn from Y 1 500 to Y 2 000 a day.

Nori farming is truly seasonal in character, requiring some 60 working days in the winter season. The Miyoshi Brothers spend the remainder of the year fishing with stationary nets in coastal waters, from which they derive about 60% of their annual income.

D.5. Sales

The farmers receive Y 12 or Y 13 for a 3 g sheet of nori of good quality. The top quality has a wholesale price of Y 200 per pack of 10. The Miyoshi Brothers produce about 300 000 sheets of nori per year. As a rule about 50% will be of top quality. Good quality nori should be glossy brown and should have a pleasant odour; the sheets should be whole and uniform. The prices paid for the poorer qualities are considerably lower. The gross income from the sale of sheets ranges from Y 3.6×10^6 to Y 4.0×10^6. In considering this figure it should be clearly understood that nori farming is a sideline, requiring only 60 working days per year and that 40% of the annual income of the Miyoshi Brothers is derived from nori, and 60% from fishing.

The total production of nori in Japan amounts to some 6×10^9 sheets, corresponding to 222 000 tons fresh nori, or 18 000 tons in terms of dry weight. The value of this product is Y 7.2×10^{10}. It is obviously a product of major economic importance among the various crops taken from the sea.

E. GOVERNMENTAL ASSISTANCE

The government owns the grounds on which the nori farmers build their racks and moore their rafts. It is by the mediation of the cooperative association of which the farmers have to be members, that they acquire the ground they need for a nominal fee.

Further, the Prefectorial Seedling Centres (or similar governmental or semi-governmental organizations), take care of the delicate first phase of growing the nori, the Conchocelis stage, for the vast majority of the farmers. This is an enormous advantage since it is more efficient by far to do this on a large scale, making use of the equipment for a great number of nets. Moreover, the centres possess a considerable amount of expert knowledge which is put to use in guiding the operations scientifically, and have easy access to scientific institutes in case of emergencies.

The farmer has to pay a fee for the seeding of his nets, but this fee does not cover the total costs of the Seedling Centre, including the wages and amortization. Thus there is a degree of subvention of this branch of the fishing industry. Concentration of effort and expert knowledge provides, moreover, for a much faster technical development than would be possible if all the work was placed in the hands of the individual farmers, who seldom have the financial means to experiment on a large scale with new systems.

In the next phase of the farming, which takes place in the field, the farmers receive all the information they need on the tides and other hydrographic factors, but also on diseases and methods of control. A great number of Japanese scientists have worked on the various phases of the biology and farming of *Porphyra*, not only in the fisheries institutes, but also in the universities. An impressive number of publications have followed from these efforts. Extension services have helped the nori farmers to gain a good understanding of the background to the apparent caprices of their crop. Of the 12 members of the extension service of the Yamaguchi Prefecture Inland Sea Marine Products Experimental Station, three are working full-time on nori farming, issuing bulletins on hydrographic and other conditions every 10 days during the height of the season.

In the last phase of nori farming, packing and distribution, the fisheries cooperatives (financially backed by the Japanese Government) do their utmost to guarantee the farmer a fair price for his product. They do not hesitate to hold back shipment when the market seems to become oversaturated, and have enough influence over central Government to persuade it to limit the nori import from countries such as Korea.

F. GENERAL IMPRESSION

Nori farming as practised in Japan is virtually the only case of large-scale application in aquaculture of methods used in agriculture and horticulture. It is a case of full-cycle farming, which does not depend in any phase on the caprices of nature — a true case of rational exploitation of a natural resource. It is interesting that the rearing of the first phase of the development of *Porphyra*, the Conchocelis phase, is now predominantly carried out ashore under controlled conditions. The consistent results are ample reward for the trouble taken, and now form the backbone of the industry, which shows not only a steep rise in production, but also a lesser degree of fluctuation in production than in bygone days, when one was more dependent on nature.

For centuries nori farming was based on locally developed empirical methods, and often had to cope with serious reverses of fortune. Credit is due to the scientists who stood at the cradle of the modern nori industry, basing new methods of farming and processing on their findings. Perhaps the most important finding, leading to a sharp increase in production, was that of the British algologist Miss Kathleen M. Drew, who discovered the role of the Conchocelis phase. This is an interesting example of how findings originating in the ivory towers of pure science can quite suddenly find application in another part of the world, resulting in greater prosperity in some particular industry. It was to express their gratitude that the Japanese nori farmers erected near Kumamoto a bronze plaque memorial to Miss Drew, honouring her for her scientific achievements (Fig. 17).

Prosperity has now arrived, and the value of the total crop of nori is impressive. In the old days the work was slow, cold and poorly paid, but now this industry is highly mechanized, the work is now comfortable and clean, and, moreover, very well paid. It still attracts young workers, whereas other fishery industries in which the work is harder and dirtier, have to cope with an increasing lack of manpower. This is the main reason why nori farming supplanted oyster farming in Japan.

The share of the government agencies in developing the nori industry was an important one. They paved the way for new methods, experimented on a large scale, and instructed the farmers through their extension services. This led to rapid mechanization, which was of benefit to the whole industry,

Fig. 17. Memorial for Miss K.M. Drew near Kumamoto. (Photo courtesy National Federation of Seaweed and Shellfish Production Cooperatives, Tokyo.)

though the investments required make small-scale operations increasingly difficult. It is interesting to compare the amount of nori produced per hectare with the yields of crops grown on the land. Three or four crops can be harvested from one net in the course of the season, and from this yield, on average, about 2 500 sheets are made. One can place 100 nets/ha, but only half of this sheet of water will then be covered by nets. Hence 1 ha can yield 750 kg of dry nori, or about 7 500 kg of the fresh product. This amounts to 750 g/m^2 of fresh product. Taking account of the fact that only half of the water-sheet is covered with nets one could say that the yield is 1 500 g/m^2 of water surface used. This is almost of the same order of magnitude as many vegetable crops grown on the land, but the financial yield from the water is more interesting, i.e., Y 3×10^6/ha or Y 300/m^2, or double this if one counts only the surface used.

There is however an important difference. The arable land should either be sufficiently fertile or fertilizer must be added by the farmer. The water where nori is grown is replenished with nutrients brought by tidal flushing from a large area. If one covered an entire bay with nets the yield would surely be poorer. The nori industry is economically one of Japan's foremost sea-based industries. The total tonnage of nori produced may not seem so impressive, but the financial value certainly is; it is more than double the proceeds from all the herring landed from the North Sea! One wonders however whether this value gives a realistic impression of the value of nori as a food. Indeed it does not. Expressed in terms of its protein content, nori is definitely too expensive as a daily food and can be compared with expensive kinds of game, meat or fish. It serves the Japanese people rather as a palatable condiment than as a substantial food. Though it may not be a food to appease hunger, it contributes materially to the welfare of the Japanese and provides employment for tens of thousands of farmers and farm workers.

Pollution of inshore water is one of the greatest threats to the future of the nori industry, though raft farming may offer some relief. It has already been shown to be possible to grow nori full-cycle on a technical scale far away from the sea, as a real bio-industry. The economics of such an enterprise will certainly be very different from those of the present method of exploitation of this resource.

G. BIBLIOGRAPHY

Anonymous, 1971. Our Seaweed Research. Joint publication No. 21 of the National Federation of Seaweed and Shellfish Production Cooperatives and the Central Conference of Shallow Sea Propagation, Tokyo, 74 pp. (in Japanese, with English summary).

Arasaki, S., 1947. Studies on the rot of *Porphyra tenera* by a *Pythium*. Bull. Jap. Soc. Sci. Fish., 13: 74—90.

Arasaki, S., 1960. A chytridean parasite of *Porphyra*. Bull. Jap. Soc. Sci. Fish., 26: 543—548 (in Japanese, with English summary).

Arasaki, S., Fujiyama, T. and Saito, Y., 1956. On the relation between the appearance of spores of the laver (*Porphyra tenera*) and the environmental conditions. II. Bull. Jap. Soc. Sci. Fish., 22: 167—171 (in Japanese, with English summary).

Batters, E.A.L., 1892. On *Conchocelis*, a new genus of perforating algae. In: G. Murray (Editor) Phyc. Mem. Pt. I., pp. 25—28.

Boney, A.D., 1965. Aspects of the biology of the seaweeds of economic importance. Adv. Mar. Biol., 3: 105—253.

Chapman, V.J., 1970. Seaweeds and Their Uses. Methuen, London, 304 pp.

Drew, K.M., 1949. Conchocelis-phase in the life-history of *Porphyra umbilicalis* (L.) Kutz. Nature (London), 164: 748—749.

Drew, K.M., 1951. Rhodophytes. In: G.M. Smith et al. (Editors), Manual of Phycology, Waltham, Mass., pp. 167—191.

Drew, K.M., 1954. Life history of *Porphyra*. Nature (London), 173: 1243.

Drew, K.M., 1956. Reproduction in the Bangiophycidae. Bot. Rev., 22: 553—611.

Furukawa, A., 1971. Outline of the Japanese Marine Aquaculture. Jap. Fish. Res. Conserv. Assoc., Tokyo, 39 pp.

Imada, O., Saito, Y. and Teramoto, K.I., 1972. Artificial culture of laver. Proc. 7th Int. Sea-weed Symp., Univ. Tokyo Press, Tokyo, pp. 358—363.

Ishio, S., Yano, T. and Nakagawa, H., 1972. Cancerous disease of *Porphyra tenera* and its causes. Proc. 7th Int. Sea-weed Symp., Univ. Tokyo Press, Tokyo, 373—376.

Iwasaki, H., 1961. The life-cycle of *Porphyra tenera* in vitro. Biol. Bull. (Woods Hole Mass.), 121: 173—187.

Iwasaki, H. and Matsudaira, C., 1963. Observations on the ecology and reproduction of free-living Conchocelis of *Porphyra tenera*. Biol. Bull. (Woods Hole Mass.), 124: 268—276 (in Japanese, with English summary).

Kurogi, M., 1953. Study of the life history of *Porphyra.* I. The germination and development of carpospores. Bull. Tohoku Reg. Fish. Res. Lab., 2: 67—103 (in Japanese, with English summary).

Kurogi, M., 1953). Study of the life history of *Porphyra.* II. On the liberation of monospores from the filamentous thallus (Conchocelis-stage) of *Porphyra tenera* Kjellm. Bull. Tohoku Reg. Fish. Res. Lab., 2: 104—108 (in Japanese, with English summary).

Kurogi, M., 1959. Influences of light on the growth and maturation of the Conchocelis thallus of *Porphyra.* I. Effect of the photoperiod on the formation of monosporangia and liberation of monospores. Bull. Tohoku Reg. Fish. Res. Lab., 15: 33—42 (in Japanese, with English summary).

Kurogi, M., 1963. Recent laver cultivation in Japan. Fish. News Int., 2: 269—274.

Kurogi, M., Akiyama, K. and Sato, S., 1962. Influence of light on the growth and maturation of the Conchocelis thallus of *Porphyra.* Bull. Tohoku Reg. Fish. Res. Lab., 20: 121—126 (in Japanese, with English summary).

Kurogi, M. and Hirano, K., 1956. Effect of water temperature on the growth, formation of monosporangia and monospore hibernation in the Conchocelis-phase of *Porphyra tenera* Kjellm. Bull. Tohoku Reg. Fish. Res. Lab., 8: 45—61.

Matsumoto, F., 1959. Studies on the effects of environmental factors on the growth of "nori" (*Porphyra tenera* Kjellm.) with special reference to the water current. J. Fac. Fish. Anim. Husb., Hiroshima Univ., 2: 249—333 (in Japanese, with English summary).

Saito, Y., 1956. A study of some environmental factors upon the development and maturity of the Conchocelis-phase of the laver *Porphyra tenera* Kjellm. Bull. Jap. Soc. Sci. Fish., 22: 21—29 (in Japanese, with English summary).

Smith, H.M., 1905. The seaweed industries of Japan. Bull. Bur. Fish. U.S., 24: 133—165.

Sundene, O., 1962. Dyrking og anvendelse av tang og tare i Japan. Blyttia 20: 105—121, 145—167.

Suto, S., 1950. Shedding, floating and fixing of the spores of *Porphyra tenera.* Bull. Jap. Soc. Sci. Fish., 16: 23—26 (in Japanese, with English summary).

Suto, S., 1971. Mariculture of Seaweeds and its Problems in Japan. Paper of the First Joint Meeting of the Agriculture Panel, U.J.N.R., Tokyo, 33 pp.

Suto, S. and Umebayashi, O., 1954. On the perforating disease in nori (*Porphyra*) culture. Bull. Jap. Soc. Sci. Fish., 19: 1176—1178 (in Japanese, with English summary).

Suto, S., Maruyama, T. and Umebayashi, O., 1954. On the shedding of spores from the Conchocelis-phase of *Porphyra tenera* cultured in the shells of bivalves. Bull. Jap. Soc. Sci. Fish., 20: 490—494 (in Japanese, with English summary).

Takeuchi, T., Matsubara, T., Shitanaka, M. and Suto, S., 1956. On the shedding of spores from the cultured Conchocelis-phase of *Porphyra tenera* set in the sea. Bull. Jap. Soc. Sci. Fish., 20: 487—489 (in Japanese, with English summary).

Takeuchi, T., Shitanaka, M., Fukahara, A. and Yamasaki, H., 1956. Studies on the ecology of the Conchocelis-phase of *Porphyra tenera* Kjellm. 3. On the resistability of the Conchocelis against unfavourable conditions. Bull. Jap. Soc. Sci. Fish., 22: 16—20 (in Japanese, with English summary).

Chapter 2

MUSSEL FARMING IN THE NETHERLANDS

Example
Johannes Zuijdweg & Sons, Yerseke, acting directors Boudewijn and Jan Zuijdweg.
Period of observation: 1963—1972.

A. BACKGROUND

A.1. General principles

Mussel farming Dutch-style is a semi-culture, in which the production of mussel seed is left entirely to nature. The public seed beds are in principle closed for fishing. Mussels which settle on the natural beds do not usually grow there satisfactorily and also usually suffer heavy losses from a variety of causes. Transfer to safer and richer grounds forms the basis of the mussel farming. The government gives permission to fish on the seed beds during short, well-defined periods. Then the seed is transferred by the farmers to their private growing plots where it is evenly scattered on the beds. Though the mussel farmer can control some predators, he is faced with a number of risks beyond his control.

According to the growth rate of the seed, the farmer can decide to distribute the mussels over a greater bed area. In some cases he will transfer the half-grown mussels to plots where he expects growth and fattening to be optimal. Natural conditions control the quality and quantity of food in the water flowing over the farming plots. Marketable mussels are fished from the plots and must undergo a cleansing procedure before they can be offered to the consumer. They can be kept in storage on the cleansing plots in order to serve the market according to demand.

A.2. Biology of the species

The mussel farmed in The Netherlands is the common blue mussel, *Mytilus edulis* L. It is a bivalve mollusc of asymmetric shape. Unlike oysters, both shells of a mussel have the same degree of cupping, but the anterior end of the shell is reduced, whereas the posterior end is enlarged. The hinge is placed at the anterior end of the shell; the posterior adductor muscle is large, the anterior small. The shell is built up of obliquely oriented bluish needles of aragonite embedded in conchyolin. A thick yellowish conchyolin periostracum covers the exterior surface of the shell of the mussel, and a whitish nacre covers the interior surface. The mussel is not immobile but can displace itself, both horizontally and vertically, with the aid of its extensible foot. A byssus gland in the foot can produce a thick liquid which hardens quickly when exuded into sea water. By ejecting this fluid through a groove which extends down the rear surface of the extended foot, the mussel can produce byssus threads with which to attach itself to a substratum. The number of threads made for attachment depends on wave motion, current conditions, the season of the year, and also on the general condition of the mussel. The mussels attach themselves not only to various solid substrata, but also to shells of other mussels, which can result in the development of a carpet of mussels.

Amongst the principal organs of the mussel are the gills, each gill consisting of two broad plates composed of a large number of parallel filaments. Ciliary hairs create a water flow through the gills. The gills strain the water to filter out food. Special sets of ciliary hairs convey the food entangled in mucus towards the mouth. Digestion takes place in the alimentary tract, and intracellular digestion in the chocolate-brown digestion gland ("liver"), situated in the centre of the body. The mussel feeds on phytoplankton, alive or dead. To collect sufficient food the mussel filters impressive quantities of sea water, the quantity increasing as water temperature, salinity, and other characteristics of the water approach the optimal levels for the mussel. Growth and fattening therefore show a seasonal pattern in the temperate and boreal waters where this mussel occurs. As in other bivalves the mantle produces shell, from the edge in order to extend the size of the shell, and from its entire surface in order to thicken the shell. Inside the mantle tissues, are situated the reproductive organs, which produce either eggs or sperm. Reproduction may take place at any time of the year, but in the Dutch mussel area spawning occurs predominantly from March to June, sometimes followed by a second peak in late summer. The mussel can produce millions of eggs, and in so doing loses a considerable amount of its reserve glycogen. Hence the mussels are so lean after spawning that they cannot be marketed. The spawned eggs are fertilised in the open water by the sperm produced by simultaneously spawning male mussels. The fertilized eggs rapidly develop into trochophora larvae, then into veligers, which lead a planktonic life, carried hither and thither by the tidal currents. When they have reached a shell length of 0.25 to 0.30 mm the veligers attach themselves with byssus threads to filamentous objects such as hydroids and certain red-weeds. They can just as easily detach themselves and either crawl around or travel on the currents, often to places far away from the initial site of attachment, to congregate with other juvenile mussels and form the often thickly-populated natural seed beds. Growth depends largely on feeding conditions at the site of final attachment. When mussel farmers transfer the seed to cultivation plots, the growth rate of the mussel can be increased considerably. The mussels are marketable in the Dutch mussel farming area when they are from 2 to 3 years old. The observation that mortality strikes heavily when fully grown mussels are kept longer than this, indicates that the life-span does not exceed more than about 4 years, at least under cultivation conditions.

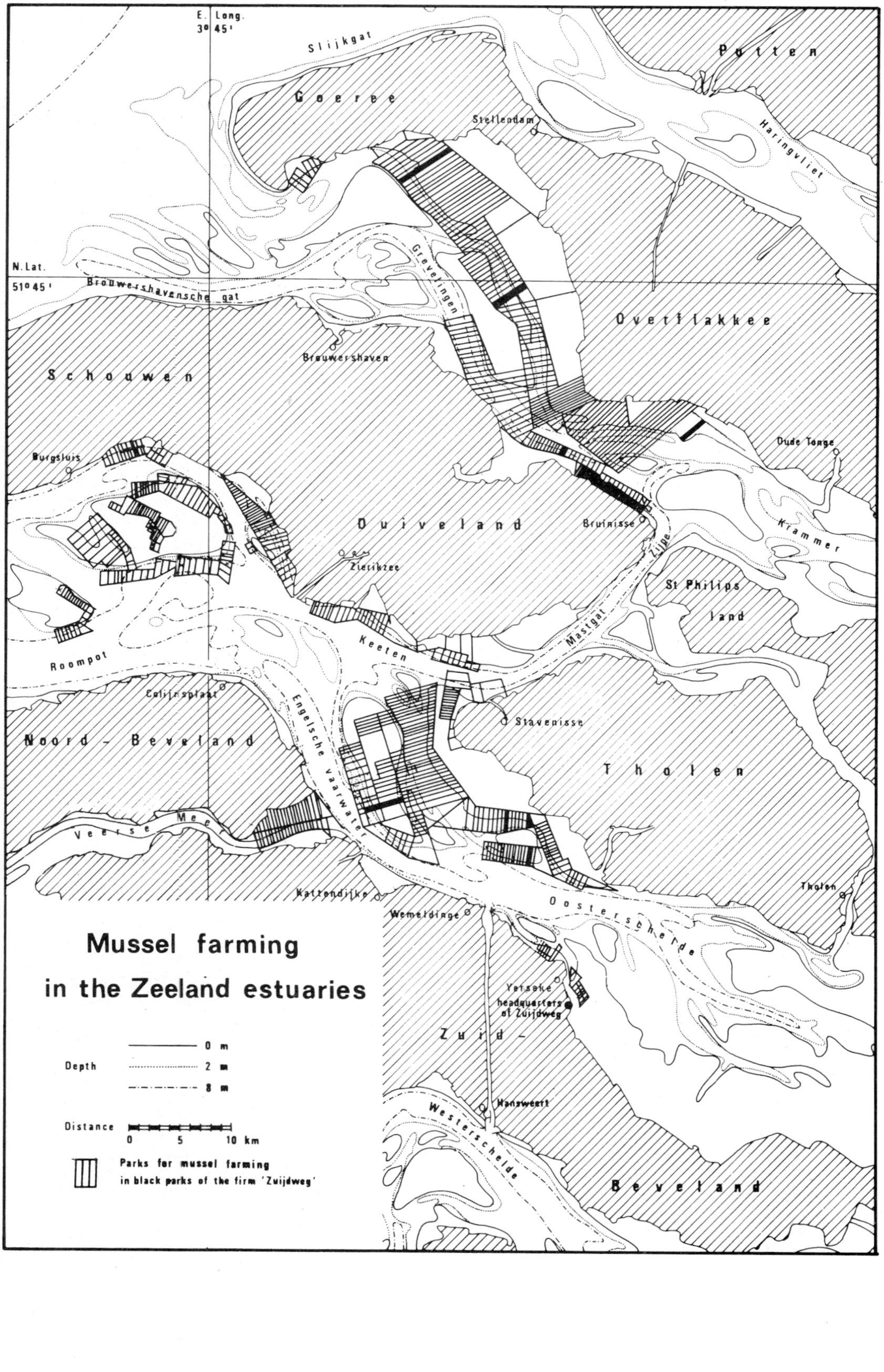

Mussel farming in the Zeeland estuaries

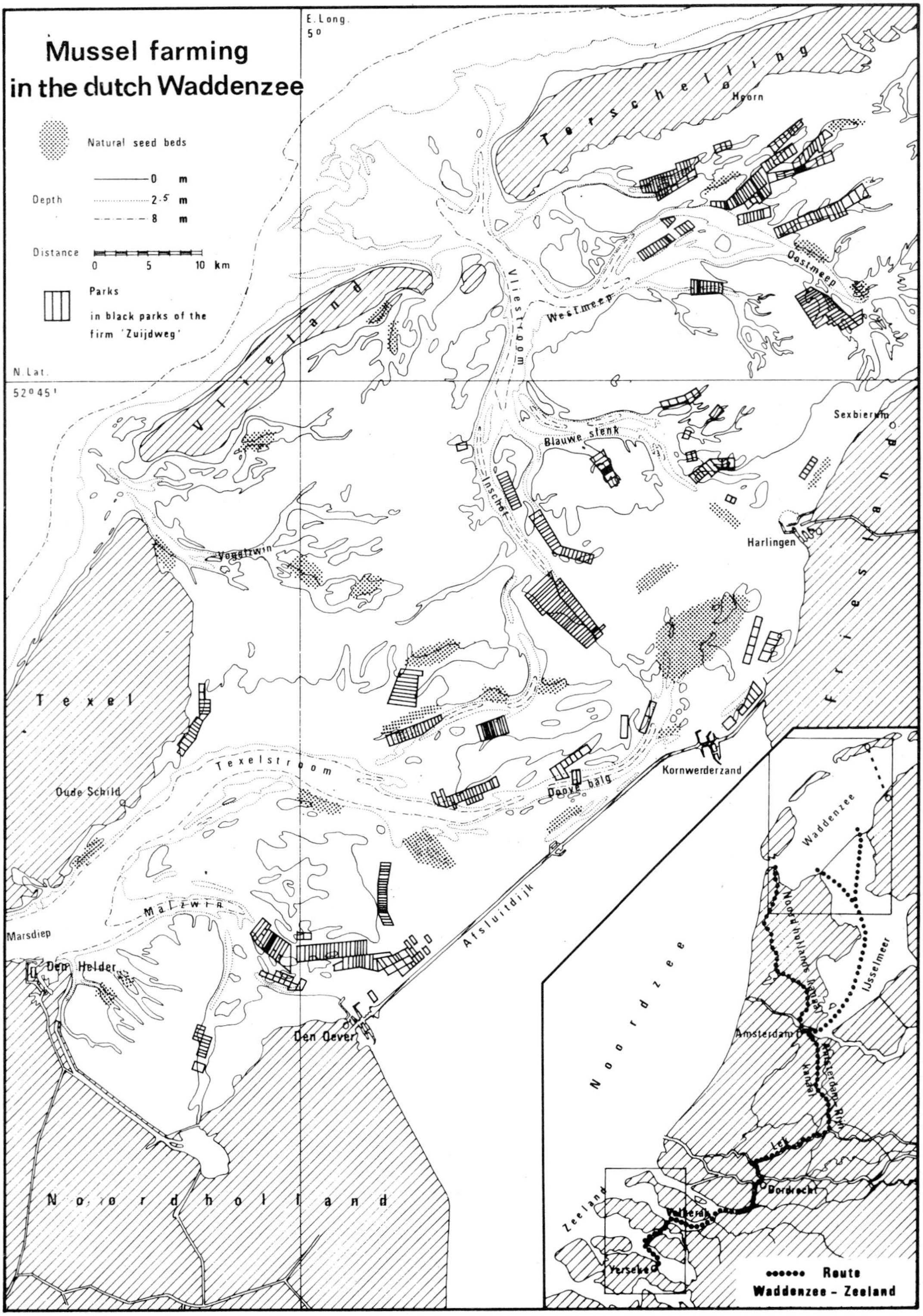

Mussel farming
in the dutch Waddenzee
Natural seed beds
Depth
0 m
2.5 m
8 m
Distance
0 5 10 km
Parks
in black parks of the firm 'Zuijdweg'
E. Long.
5°
N. Lat.
52°45'
Terschelling
Hoorn
Vlieland
Vliestroom
Westmeep
Oostmeep
Sexbierum
Blauwe slenk
Inschot
Harlingen
Vogelzwin
Friesland
Texel
Texelstroom
Oude Schild
Doove balg
Kornwerderzand
Malzwin
Marsdiep
Den Helder
Afsluitdijk
Den Oever
Noordholland
Waddenzee
Noordzee
Noordhollands kanaal
IJsselmeer
Amsterdam
Amsterdam Rijn kanaal
Lek
Dordrecht
Zeeland
Yerseke
Route
Waddenzee - Zeeland

A.3. Geographical situation

Formerly the Dutch mussel farming plots were exclusively situated in some sea inlets in the southwest of The Netherlands; the Waddenzee in the northern part of the country was used as a source of seed. Since 1950 this pattern has changed considerably; farming plots were created in the Waddenzee too, and now from 60 to 70% of the marketable mussels are harvested in the Waddenzee, the remainder being produced on the old beds in Zeeland. The company under consideration, having actively participated in this development, therefore has plots both in Zeeland and in the Waddenzee (see Charts).

The headquarters of the company are situated 51°29′ N and 4°02′ E in the town of Yerseke (Zeeland). All the mussels marketed by the company are taken there after a short spell on the cleansing and storage plots near Yerseke.

In Zeeland the more important of the mussel plots are to be found in those sections of the Oosterschelde situated west of the town of Yerseke. The Oosterschelde is an inlet penetrating about 48 km inland from the North Sea to the city of Bergen-op-Zoom, i.e., approximately from 3°40′ E to 4°15′ E, in a direction WNW—ESE. To the south one finds, going from west to east, the islands Walcheren, Noord-Beveland and Zuid-Beveland, and to the north are the islands Schouwen-Duiveland and Tholen. The narrowest sections, off Zierikzee and off Yerseke, are about 4 km wide. Due north of the island of Schouwen-Duiveland is the inlet called Grevelingen, in which other groups of mussel plots are located.

The Waddenzee plots are all situated in its western section, in the area approximately defined by the points 52°55′ and 53°23′ N and 4°15′ and 5°30′ E. It measures about 70 km in a direction SW—NE and has a maximum width of about 30 km. It is continuous with the North Sea through the wide Marsdiep in the southwest, and through minor passages between the Isles of Texel, Vlieland, Terschelling and Ameland. The company's ship operating in the Waddenzee has to make a fair number of voyages to Yerseke, for instance to transport marketable mussels there. It never takes the dangerous route by the North Sea, but follows the water-ways in the interior of the country, usually via the IJsselmeer (passing either the locks of Den Oever or of Kornwerderzand), through the locks at Amsterdam, and takes the big canal southward to the river Rhine. It follows the Rhine downstream, and after passing the city of Dordrecht, it has only to negotiate the Volkerak locks to sail unhindered to Yerseke.

A.4. Hydrographic pattern

I. Oosterschelde

Depths and dimensions. The Oosterschelde is an inlet noted for its deep channels (15 to 35 m depth) and its extensive tidal flats. Its average depth at high tide is 12.20 m. The section west of Yerseke measures about 22 000 ha of water at high tide, but only 17 000 ha at low tide, so the tidal flats therefore take up about 23% of the total area. Mussel plots are never laid out in the deep channels which have strong tidal currents and beds of shifting sand, nor on the tidal flats, where the mussels would be exposed for many hours a day. The Zeeland mussel plots are to be found from the low water level down to a depth of 5 or 6 m. Their official outline on the charts may include some higher and some deeper grounds, but the mussel farmers use only those parts of their plots which they deem suitable for farming.

Tides. The tidal range varies from 3.40 m at spring tide to 2.70 m at neaps, with an average of about 3 m. This involves a tidal displacement of about 600×10^6 m^3 of water, for the section west of Yerseke amounts to 2 450 $\times$ 10^6 at high tide and 1850 $\times$ 10^6 m^3 at low tide. This displacement results in strong tidal currents: in the deeper channels about 200 cm/s at spring tide and 150 cm/s at neaps; in the other channels 170 cm/s and 130 cm/s respectively. Lower velocities prevail on the mussel plots.

Waves. The height of the waves is dependent on wind velocity, water depth, and the degree of shelter. In the western-most section waves 2 or 3 m high are common during stormy weather. In areas sheltered by the islands the waves are less high, and the cleansing and storage plots near Yerseke are perfectly sheltered against western winds.

Water temperature. There is a marked seasonal pattern in water temperature. In winter, the temperature is predominantly between 0 and 5°C. Once in four or five winters ice may form in the Oosterschelde, indicating a water temperature as low as —1.5°C. The tidal currents keep the drift ice in motion. Such conditions rarely last longer than about 4 weeks. In spring, temperatures rise quite rapidly, reaching 10°C in April and exceeding 15°C by about the middle of May. Summer maxima may reach between 19 and 21°C, but occasionally are as low as 18°C. Towards the first of October the water temperature will have dropped below 15°C, and one month later, is about 10°C. Due to the strong tidal currents stratification does not occur. The mussels display the greatest activity at temperatures over 10°C.

Salinity. The salinity of the Oosterschelde is remarkably constant, remaining within the range 27 to 32‰, average about 29‰. Influxes of fresh water are insignificant in this area. The salinity of the Oosterschelde water is therefore virtually the same as that prevailing in the continental coastal water, which is influenced by variations in the run-off from the river Rhine.

Plankton. The phytoplankton reveals marked fluctuations in both qualitative and quantitative aspects. During the spring bloom there may be 50—100 mg chlorophyll per m^3, during the secondary bloom late in summer, 30 to 40 mg/m^3, whereas in winter one may record only 10—20 mg/m^3. The species succession pattern of diatoms and flagellates may be surprisingly rapid. A high chlorophyll figure does not necessarily mean good feeding conditions for mussels, for it may coincide with temperatures so low that the mussels are able to filter only very slowly, and it may be moreover that very spiny diatoms or bulky species such as *Phaeocystis*, neither of which are favourite food organisms for mussels, are the main contributors to the high chlorophyll values.

Silt. The silt content of the water of the Oosterschelde is quite low, with a marked seasonal pattern. Of course, wind exerts a considerable influence on the amount of silt suspended in the water. In winter, one may find in the western section about 20 mg/l, in summer, as little as 10 mg/l.

Oxygen. Due to thorough tidal mixing and the virtual absence of pollution in the section of the Oosterschelde west of Yerseke, the quantity of oxygen dissolved in the water is nearly always close to the saturation value. There is a slight daily fluctuation with a peak in the afternoon and a minimum shortly before sunrise, but the range is surprisingly narrow.

Pollution. The water of the Oosterschelde is largely uncontaminated since no sewers of any great importance discharge into it. The town of Yerseke, from which considerable quantities of mussels and oysters are shipped, does not discharge any waste at all into the Oosterschelde, and has brought in municipal legislation to prevent any form of pollution in the shellfish area.

Sea bed conditions. The Oosterschelde shows a remarkably wide variation in sea bed deposits. In the channels one often finds quite coarse shifting sand, and in some deep pits, a sandy mud; locally the bed consists of clay deposits or peat soil, the latter recalling a time when these parts were not yet invaded by the sea, but the mussel plots usually have bed deposits of a stable mixture of sand and mud in which a foot does not easily sink. Mussels may deposit soft silt on top of the beds on account of their filtering activity, but by fishing the mussels with the heavy dredges used in this area, most of this soft silt is washed away again.

The hydrographic pattern of the Grevelingen deviates so little from that of the Oosterschelde that no separate description need be given. In exceptional years, in periods of excessive outflow from the large rivers Rhine and Meuse, a surface layer of fresh water over the sea water has been observed. This can cause damage to mussels grown on higher grounds, especially in the eastern part of the Grevelingen.

II. Waddenzee

Depth and dimensions. The average depth of the section of the Waddenzee west of the Isle of Ameland is only about 3 or 4 m, considerably less than that in the Oosterschelde, but the channels leading towards the open sea are quite deep, e.g., Marsdiep 25—30 m, Vliestroom as much as 30—40 m, and channels in the more central part of the Waddenzee, 10 to 15 m. This section of the Waddenzee has a total area of 158 000 ha of which 35% are exposed at low tide. The mussel plots, totalling about 4 000 ha are not located in the deeper channels, but are found in places where the depth is only a few metres. They are often laid out across secondary channels, reaching up high on the tidal flats, but the mussel farmers will only use the deeper parts of these plots.

Tides. The tidal range in the Waddenzee is considerably smaller than that in the Oosterschelde: at Den Helder it is from 150 to 118 cm, at Harlingen from 198 to 166 cm, and at Terschelling from 189 to 161 cm, spring and neap tides respectively. This section of the Waddenzee includes about 6.5×10^9 m^3 of water at high tide, compared with 4.5×10^9 m^3 at low tide. Hence the tidal displacement is in the order of 2×10^9 m^3, which calls for quite strong tidal currents. In the Marsdiep, current velocities up to 220 cm/s have been measured, in the Vliestroom 135 cm/s, and in the narrow channels 50 to 100 cm/s, whereas one observes velocities of only 20 to 30 cm/s over the tidal flats.

Waves. Though the Waddenzee is less sheltered than the Oosterschelde, wave heights exceeding 3 m are rarely observed.

Water temperature. The range of water temperatures in the western section of the Waddenzee does not differ noticeably from that in the Oosterschelde; it may be about 1°C lower on average, which means that ice forms earlier here than in the Oosterschelde, and that temperatures over 20°C occur only rarely and locally.

Salinity. The salinity in the western section of the Waddenzee fluctuates somewhat more than that in the Oosterschelde and shows a slightly lower average; the range is from 22 to 32‰,with an average of 28‰. When large volumes of fresh water are sluiced out at Den Oever, Kornwerderzand or Harlingen, these may reduce the salinity over quite an extensive area, resulting in salinities as low as 10‰. Some of the mussel plots may be affected by such low salinities.

Plankton. The phytoplankton here also shows considerable seasonal fluctuations, both qualitatively and quantitatively. During the spring bloom one may find 100—150 mg chlorophyll per m^3, and during the peak late in the summer up to 200 or 250 mg/m^3, whereas in winter and in the period May—June, there may be as little as 20—25 mg/m^3.

Silt. The silt content in the Waddenzee is noticeably higher than that in the Oosterschelde. Wind, currents and depth affect the quantity of suspended silt to a considerable extent. It is rarely less than 10 mg/l, fluctuating in winter around 50 mg/l, and in summer around 25 mg/l, but in silt clouds swirled up by the tides and the wind one can find 200 to 400 mg silt/l.

Oxygen. The daily fluctuation in the quantities of dissolved oxygen are, as a rule, not very great, with minima of 90 to 95% saturation. In the water layer immediately above a dense population of mussels, values around 80% saturation have been measured, with minima occurring during the night.

Pollution. The Waddenzee can be considered as a largely unpolluted area with some local contamination due to an influx of domestic sewage near the cities of Den Helder and Harlingen. Therefore the mussels produced here can meet the international standard of bacteriological safety. Though the reduction in the salinity (relative to that of the open water of the North Sea) is mainly brought about by Rhine water which arrives either via the IJsselmeer or by skirting the mainland, no proof can be given that the mussels here suffer from industrial pollution of any description.

Sea bed conditions. There is a considerable range in the particle size of the sea bed deposits, from soft silt of very small particle size to coarse sand. Mussel plots are found in areas where a mixture of sand and silt predominates.

A.5. Legal aspects

The company under consideration was established as a private enterprise by Johannes Zuijdweg and is at present run by his sons Boudewijn and Jan.

The grounds on which mussels and oysters are farmed belong almost exclusively to the State, both in the Zeeland waters and in the Waddenzee. The Finance Department and more specifically its section Crown-lands ("domeinen") can lease plots to mussel farmers and oyster farmers. The Zeeland areas suitable for farming have been established over many years and it is most unlikely that groups of plots on hitherto barren ground can be created. After a long historical development, starting in the nineteenth century, the original system of auctioning the plots to the highest bidder, when the official 11-year lease has elapsed, has been abolished. Nowadays the lessees are fairly well protected and will not lose their plots as long as it can be demonstrated that they exploit the plots efficiently and in keeping with the actual magnitude of their enterprise. The rents differ from plot to plot, the rate being related to the historical development rather than to the practical results recently obtained from the plots. The rents are quite low and are never a heavy burden on the budget of the shellfish farmers. A new plot can only be acquired when another company abandons its plot for some reason, and when a real need of more ground can be demonstrated. In the Waddenzee, where farming started in 1950 and where the rights of shrimp fishermen must be taken into account, it is still potentially possible to create more plots. The rent paid per hectare is uniform there.

A special booklet ("Algemeene Voorwaarden van Verpachting") gives an explanation of the rights and duties of the lessees, the more important of which are first, the duty to place seamarks in order to delineate the plots, and secondly, to keep the plots in good condition so that neighbours will not be troubled by predators or competitors spreading from ill-kept plots. Disregarding a warning about these matters results in loss of the plot. The company under consideration farms nine plots in the Waddenzee and eight plots in Zeeland (see Charts).

The fishery inspectors are responsible for the correct staking out of the plots. On higher ground, markers of creosoted oak poles provided with copper number plates are used at the four corners of each plot. Such poles do not get lost in winter. A simple reflecting prism and a sextant are used to establish the angles and a 4 mm steel wire provided with marks at known distances serves to measure the breadth of the plot. The mussel farmer has to place more seamarks at the seaward border of the plot and should refund the inspector for the expense of the creosoted corner poles. As seamarks the farmers use oak poles from 7 to 13 m long, inserted at 50 to 60 m intervals. The Waddenzee plots measure on average 500 × 200 m, the plots in Zeeland 300 × 150 m. Where the ground is hard, a pump and hose are used to drive the poles and seamarks into the beds.

Since the early days of leasing, the government provided for adequate police patrols to prevent theft of mussels and oysters from the plots and to enforce all the rules which apply to the lessees, even when in due course the cost of this surveillance and the use of boats tended to exceed the income from the rents. There are still several police boats operating in the mussel area, but particularly in the Waddenzee the distances are so great that it would be very costly indeed to organize a very intensive patrol covering all groups of plots adequately. Mussel farmers claim that some theft does occur, and that there is fraudulent displacement of seamarks, which can easily go unobserved by the police officers in charge during the actual fishing operations.

B. SEQUENCE OF OPERATIONS

B.1. Procuring the seed

Mussel seed settles in various places which provide a suitable substratum, both in the intertidal zone and in deeper waters. Dense populations of juvenile mussels are often found in approximately the same places year after year. Such places are referred to as the natural seed beds. In The Netherlands by far the richest natural seed beds are to be found in the Waddenzee, in the lee of the Friesian Islands. Natural beds in the Zeeland waters on the other hand, cannot rely on a good regular settlement of mussel seed. Therefore commercially sound mussel farming cannot be based on the use of seed solely from the Zeeland beds. A third site where profuse settlement of mussel seed occurs is formed by the breakwaters which have been constructed in several sectors of the Dutch and Belgian coast. This latter type of seed is usually strong and thick-shelled, excellent for farming, but difficult and costly to harvest. In the years following 1960 this latter type of seed could no longer be purchased by the Dutch mussel farmers.

The sea bed both in the Waddenzee and in the Zeeland estuaries is Government property and as such is administered by the section Crown-lands ("domeinen") of the Finance Department. The Ministry of Agriculture and Fisheries can request the Crown-lands section to open the public beds for fishing within a well-defined period. The Ministry acts on advice from its officials patrolling aboard inspection ships in the area under consideration. The officers on these ships have a thorough knowledge of their district, are daily in

Fig. 18. Mussel seed on soft-bottomed intertidal ground in the Waddenzee. (Photo A.C. Drinkwaard.)

contact with fishermen, and actively search for mussel seed in the appropriate season. They inspect buoys and seamarks, and frequently take pieces of hydroids and red-weeds between their fingers or teeth to determine whether the very small mussel seed has settled. The seed may be only the size of a sand grain to be detectable, but the inspectors will not advise the Ministry to open the beds before the seed is large and strong enough to be fished and transferred to the farming plots of the fishermen. The best size for this is about the same size as a grain of wheat up to that of a haricot bean (Figs 18 and 19). The officers realise that by waiting too long before opening the beds, green algae may grow so profusely on the seed that harvesting becomes difficult; delay may also give starfish more opportunity to devour the seed, and the strong currents of the spring tides may destroy whole beds of mussel seed in some exposed sites. Water temperature is also a factor to be taken into account, for at lower temperatures the seed is often difficult to fish, being too strongly attached to the substratum. When the officers have advised the Director of Fisheries as to the most appropriate time to open the public beds, the latter contacts the mussel farmers' organisations to select in consultation with them an appropriate date for opening the beds, taking into account the other seasonal work of transferring half-grown mussels to other

Fig. 19. Healthy mussel seed, to be transferred to the farming plot, is well-clustered. (Photo A.C. Drinkwaard.)

plots, and possible interruption by week-ends and Church holidays. Usually the date selected for the Waddenzee beds is in late spring or early summer. As a rule all the beds in the western Waddenzee, west of the line Sexbierum (Friesland)—Hoorn (Isle of Terschelling), are opened for fishing simultaneously for 2 or 3 weeks, starting at 7 a.m. on the first morning. Fishing is allowed during daylight only.

The quantity of mussel seed available on the natural beds in the western Waddenzee is usually sufficient to stock the farmers' plots, but occasionally there is a shortage. If that is the case one searches the natural beds for mus-

Fig. 20. Fishing mussel seed on the Waddenzee with a modern ship equipped with four dredges. When many ships participate in fishing on a limited acreage manoeuvring is no easy matter. (Photo L. Westbroek.)

sels one year older with which to start the cultivation, though these mussels may not always be of the desired quality. Some farmers try to create a reserve supply by putting some of the fished seed on a plot where growth is slow. This was formerly common practice in Zeeland, and special plots were rented for that purpose. In the Waddenzee there is often a shortage of available plots, and the risks of losing the seed in winter are greater there than in Zeeland. Planting the seed too densely on the plots is common in years when there is an abundant supply, but growth is usually poor on over-populated grounds, and mortality is high.

Gathering the mussel seed. Nowadays gathering mussel seed is done exclusively by special mussel boats [1*] (Fig. 20) equipped with diesel-engines and either two, or in most cases four, powered dredges [2]. Each dredge can be operated individually, for which purpose the vessel is equipped with a special type of winch[3]. One dredge load is from 400 to 500 kg of mussels. When the

* The reference numbers refer to paragraphs giving more detailed description in the section "Technical details" in this and following chapters.

beds are well supplied with seed the dredge is filled after 10 min towing. This means that for a boat equipped with four dredges, a dredge has to be hoisted every 2.5 min. Before the filled dredge is swung inboard, the operator in the wheel-house dips it rapidly and repeatedly in the sea to get rid of most of the adhering silt. Once over the deck, the dredge is lowered and lifted from behind with the aid of a special cable from the winch, in order to empty it into the hold. Thus about 10 000 kg of mussel seed can be collected after dredging for 1 h; when the bed is exceptionally well-supplied one can harvest 15 000 kg or more per h. The hold of Mr. Zuijdweg's ship can take from 50 000 to 60 000 kg of mussels and it can be filled in 3 to 5 h fishing. To carry out this work efficiently the boat should have a crew of at least three, or if possible four, one in the wheel-house operating the wheel and the winches, the other three on deck and in the hold, to swing the dredges inboard and outboard, to empty the dredges into the hold taking care that the boat is evenly loaded to avoid listing, and also to pick out by hand most of the starfish found in with the mussels. The latter job is necessary only when the starfish are a fair size. If numerous small starfish are found among the seed they are killed by sprinkling with salt. For this purpose one sack (50 kg) of fine grained tax-free salt per 10 000 kg of mussel seed is used, and the salt is added gradually during the fishing operation, after an initial 10 000 kg of seed have been hoisted into the hold. This treatment entails a certain risk for the mussel seed, especially when it is extremely thin-shelled. In any case, mussel seed treated with salt should be planted the same day and should not be kept overnight in the hold.

The home-port of the ship of the firm under consideration is Yerseke (Zeeland). The journey to the Waddenzee via Volkerak, Dordrecht, Vreeswijk, Amsterdam and the IJsselmeer, takes about 24 h. During seed fishing operations the ship stays overnight in the port of Den Oever, Harlingen or Terschelling. The mussel seed fished on the natural beds of the western Waddenzee should be used primarily to supply the growing plots in the Waddenzee itself. Only when there is a very plentiful supply, combined with serious shortages elsewhere, does the Ministry give permission to transfer some of the Waddenzee seed to growing plots in Zeeland. The Zeeland plots are normally supplied with seed either produced in Zeeland (from natural beds and breakwaters; some Zeeland plots on which natural settling occurs now and then are leased for this reason) or from the eastern section of the Waddenzee, east of the Isle of Ameland. The mussel seed from this eastern section may on no account be transferred to the plots in the western Waddenzee, since it contains the parasitic copepod *Mytilicola intestinalis.* The plots in the western section of the Waddenzee are still free of this parasite. Transfer to Zeeland is permitted on the other hand, because the Zeeland estuaries have been invaded by this parasite since 1949. The presence of *M. intestinalis* does not put the mussel consumer at risk but, if present in larger numbers, it may inhibit growth, resulting in mussels in poor condition, and in serious cases it may cause the death of the mussels.

B.2. Planting the seed

The seed is planted on one of the growing plots the farmer has leased from the government. The firm of Zuijdweg has nine plots at its disposal in the Waddenzee. Each plot [4] measures 500 × 200 m, is usually set out crosswise over a channel, but is not suitable for mussel farming over its entire surface. On average, mussels are distributed over about half of the area of each plot. The mussel farmer has to judge which of his plots is the best suited for planting the seed, and which section of that plot promises the best results. Apart from factors such as depth, current velocity and salinity, it is the nature of the sea bed which is the main criterion. Submerged grounds are probed with a long pole with a wooden knob at one end, an iron point at the other, both

Fig. 21. When the hold is filled to the brim with mussel seed, the ship proceeds to the parks where the seed will be planted. The dredges used by the Dutch mussel farmers are clearly shown in this picture. (Photo L. Westbroek.)

of which are used alternately. The nature of the impact and the kind of material sticking to the iron pin provide information about the character of the bed. Coarse sand and soft mud are easily recognised and avoided, but it takes a lot of practical experience to detect more subtle differences. A study of what the dredge brings up may give some further information on the features of the bed. Experience gained in Zeeland with its deeper plots and greater tidal range does not always hold good for the Waddenzee, and probing should be repeated regularly if there are reasons to suppose that the local bed conditions might change with time. Gales, and the general tendency of the tidal currents in the western Waddenzee to move in a SW—NE direction, compared with the NW—SE currents prevailing prior to the enclosure of the Zuiderzee, are factors to be taken into consideration.

Another important decision of the mussel farmer is the quantity of seed to be planted. Practical experience has shown that no more than about 8 kg of marketable mussels can be harvested per m^2 of mussel plot, or in practical terms, 800 mussel tons (tons of 100 kg) per ha. The size of the seed should be taken into consideration in deciding the quantity to be planted. When the seed is the size of wheat grain, for example, about 200 mussel tons/ha are sufficient, and if it is the size of haricot beans one should plant from 300 to 350 tons. When planting the seed, the skipper manoeuvres his ship slowly and systematically over the selected plot while the mussel seed is discharged overboard. This operation would require a considerable amount of heavy manual labour if special mechanisms were not devised to reduce it. Steadily increasing wages and the growing aversion to hard toil led to rapid development in this field. Three different systems [5] were used by the firm of Zuijdweg during the period under consideration. It is now the conveyor belt system which is used universally by the Dutch mussel farmers as this limits manual labour to an absolute minimum. The hold of the ship, containing from 50 000 to 60 000 kg of mussels is mechanically emptied and the mussels are distributed over the plot in 50 to 60 min (Figs 21 and 22).

B.3. Operations during the growing period

When the young mussels have been planted as evenly as possible, it is expected that by their own activity they will in due course form a mussel carpet in which every individual mussel can filter water in search of food. During the period of growth, the length of which is governed by natural conditions in the supernatant water, the plot should be inspected at least every other month, but less frequently in winter. Inspection is carried out by making very short dredge hauls and by probing with the pole. Thus the mussel farmer can find out whether the mussels on his plot show signs of over-population, which is indicated when the mussels form ridges and small domes. At no time should there be substantially more than 8 kg of mussels per m^2 on the beds. If the plot is too heavily stocked, the farmer distributes the mussels over a larger

Fig. 22. Planting mussel seed on a plot, making use of vertical chutes. Now that the hold is practically empty the higher section of the chutes have been taken away; one of those can be seen in the upper part of the picture. (Photo A.C. Drinkwaard.)

area by removing some of the mussels and relaying them elsewhere.

Another sign of danger is the presence of a large number of starfish (*Asterias rubens*). If the starfish are very small, the best remedy is to dredge up the mussels and give them a salt treatment as described above, or to keep the mussels overnight in the hold of the ship. When larger starfish abound it pays to work the plot with special starfish nets [6]. Four starfish nets preceded by wooden rollers are operated simultaneously. These remove the starfish without disturbing the mussels.

Sometimes the mussels produce a lot of silt from their filtering activity, and since this silt is deposited under the mussel carpet, it may be a cause of concern to the farmer. On the one hand, silt deposition may reduce both the depth of the water and the current velocity, thereby restricting the food supply, and on the other hand, a mussel carpet on a thick layer of soft mud is liable to erosion by strong tidal movements. The farmer may therefore decide to take up all the mussels and level the ground with a chain-harrow, measuring 4 × 4 m and weighing about 1 500 kg. This work is best done at a spring tide, and if possible a whole group of plots should be cleared together, since much silt is stirred up. Harrowing is faster than the former practice of dredging with bagless dredges.

In winter, growth slows down, starfish activity diminishes, and the mussels have a greater tendency to cluster and dig themselves in to some extent. Therefore less frequent inspection suffices during the cold months of the year.

The following spring the frequency of inspection is increased again in order to determine whether starfish collection, thinning out, or ground levelling are required. The mussels will measure 4 to 4.5 cm in shell-length by now, and usually reach marketable size the same year.

In the Waddenzee the mussels are often kept on the same plot till they reach the marketable size. In Zeeland, on the other hand, it is common practice to transplant the half-grown mussels to deeper plots where growth and fattening are optimal. The firm of Zuijdweg works nine plots in Zeeland (three of which were lost in 1971 on account of the Delta-project in the Grevelingen), some of which are particularly suitable for smaller mussels, others for the production of marketable mussels. The advantage of working plots both in Zeeland and in the Waddenzee lies in the spreading of risks and better adaptation to market requirements. Waddenzee mussels are as a rule larger, and therefore especially suitable for raw consumption on the half shell ("moules parquées"), whereas Zeeland mussels often reach top quality somewhat earlier in the season and because they are preferred as raw material for the canning factories, often fetch a slightly higher price.

B.4. Harvesting marketable mussels

The Industrial Board for Fisheries fixes the dates on which the season for the sale of marketable mussels will be opened and closed; in this matter it is counselled by its Mussel Advisory Board. In recent years, the season has always been officially opened on the 1st of July, and closed in April, two weeks after Easter. The quality of the mussels determines the demand, which usually reaches a peak from September to November. From the 1st of July the mussel farmer inspects his crop regularly to see whether his mussels meet the requirements for marketable mussels set by the Industrial Board. The minimum size is 5.5 cm shell-length, but since it is impossible to separate a few smaller specimens, at least 50% by number of the mussels offered for sale should measure 5.5 cm and over, about 30% from 5.0 to 5.5 cm, not more than 10% from 4.5 to 5.0 cm, and another 10% from 3.0 to 4.5 cm; mussels under 3.0 cm are considered as tare. This corresponds to a maximum of 160 mussels over 3.0 cm in a sample of 2.5 kg. Moreover, the edible parts ("the fish") should be not less than 16% of the total weight. High quality mussels measure between 6.0 and 7.0 cm shell-length and contain 25 to 30% flesh, by weight. It is the farmer's decision when to offer his mussels for sale. Since mussels of greater average shell-length and a higher percentage of edible flesh will fetch a higher price in the auction, he has not only to take into consideration the current demand, but also what further improvement in quality he could expect were he to leave the mussels some weeks or months longer on the plots. When the farmer decides to sell his product, he applies for a transportation licence to the Industrial Board and fishes his mussels by dredging.

The boat of the firm of Zuijdweg, the same one that is used for fishing the seed, can collect 10 000 kg of marketable mussels per hour. If the weather prospects are such that it will be safer to transport the mussels under the hatches, the ship can hold a maximum of 56 000 kg of mussels. Once harvested, the mussels should be transported as quickly as possible to Yerseke, where they are first sold, and then transferred immediately to the cleansing plots. If the mussels are taken from a plot in the Waddenzee, transport will take about 24 h. Where the ship has to pass locks it will take priority over empty ships and ships loaded with non-perishable goods. Mussels harvested from plots in Zeeland take at most a few hours to reach the port of Yerseke.

Considering the distance from Yerseke to Waddenzee it will be clear that a ship can make only one trip per week to the Waddenzee to fish marketable mussels. The old arrangement was to sail from the port of Yerseke after midnight on Sunday night, and work on the plots in the Waddenzee on Tuesday and Wednesday, to return on Thursday with a load of marketable mussels. The mussels were sold on Friday, and were planted on the cleansing plots as soon as possible afterwards, depending on the time of high water. The introduction of the 5-day week in the mid-nineteen sixties led to a considerable reduction in the time available for working the plots in the Waddenzee. If there is an overload of work it is necessary to hire additional ships more often than in previous years. If, on the other hand, the work on the plots is to a certain extent neglected, unavoidable losses due to various factors may increase.

B.5. Sale of marketable mussels

The mussel farmer has to offer his marketable mussels for sale at the Yerseke auction, even if he is as much a mussel dealer as a mussel farmer. There are about 120 mussel farmers in The Netherlands, and only 12 mussel dealers. Mr. Zuijdweg is both a farmer and a dealer.

Upon arrival at the port of Yerseke an official of the Industrial Board of

Fisheries determines how many tons of mussels the hold of the ship contains, using a gauging-rod and a conversion table. Next he takes a 2.5 kg sample with a small power grab, which enables him to take the sample from any position or depth in the hold. In his office the sample is weighed and the percentage of tare (mussels under 3.0 cm shell-length and everything other than mussels) is determined. The mussels are then boiled by a standard procedure and the edible parts taken out and weighed. The figure for the meat content is calculated without including the tare. All the measurements of total tonnage, size of the mussels, percentage of tare, and meat percentage are put at the disposal of the prospective buyers in the auction. The prices fetched at the auction fluctuate with the quality of the mussels offered and the seasonal demand of the market. Buyers are the dealers/exporters and the local mussel canneries, which take about 12% of the total mussel production, usually the top quality mussels. In the season 1969—1970 the price paid at the auction fluctuated between Dfl. 18 and Dfl. 25 per 100 kg. Peak prices up to Dfl. 45 per 100 kg are paid occasionally when there is a shortage. Dfl. 0.66 per 100 kg is charged by the Industrial Board for their work during this phase, half of this being paid by the farmer, the other half by the dealer. The Industrial Board sets a minimum price for marketable mussels; for the season 1971—1972 the minimum price was Dfl. 13 per 100 kg. If that price is not offered at the auction, the mussels are not sold but taken over by the Board, which pays the producer the minimum price, viz., Dfl. 13 per 100 kg. The Board pays for the mussels taken out of the market from a fund built up by the mussel farmers themselves by means of a tax levied by the Board on every 100 kg of mussels offered for sale. On the advice of the Mussel Advisory Board the amount of the tax can be adapted to the needs of the fund. In the season 1971—1972, when there was a high production of mussels, this tax fluctuated between Dfl. 2 and Dfl. 6 per 100 kg of mussels. The Board can sell these mussels (which are temporarily stored on some special plots in the vicinity of Yerseke), on a special auction day fixed in agreement with the Advisory Board. They are sold to both farmers and dealers, at a price at least as high as the minimum initially paid.

In the years under consideration the firm of Zuijdweg offered for sale the following quantities of mussels:

Season	Waddenzee plots (kg)	Zeeland plots (kg)
1963—1964	959 200	1 269 800
1964—1965	1 308 500	585 300
1965—1966	819 700	1 261 300
1966—1967	223 900	844 000
1967—1968	1 305 800	678 600
1968—1969	862 400	1 133 400
1969—1970	688 300	562 400
1970—1971	1 333 400	1 210 300

C. FARMING RISKS

The fluctuations in the production of marketable mussels are evidently considerable and often beyond the control of the farmer. Mussel farming involves certain risks which influence the quantities produced.

C.1. Shortage of mussel seed

Since mussel farming Dutch-style is a semi-culture which depends on nature for the production of seed, it may happen that there is not enough seed on the natural beds to stock the plots. This situation occurs about once in 5 years. The farmers then try to find older mussels on the natural beds to compensate for the shortage. Anxiety about possible shortages of seed in coming years often leads the farmer to plant too much seed in years when production is high. But is does not pay to cover the plots with a layer of seed which is too thick.

Any settling of mussel seed on the mussels already growing on the plots is most unwelcome. If the larger mussels are nearly ready for the market, the presence of seed, even in small quantities, increases the percentage of tare, and thus reduces the price. The dealer who buys such mussels has to remove the seed from the marketable mussels before he can sell them to the consumer. If mussel seed settles in very large quantities, the older mussels are usually smothered, and are progressively killed. The seed itself may remain viable, but it is not always of sufficiently high quality for farming. If the farmer notices that seed has settled profusely on his mussels he can immediately gather the entire crop and keep it overnight in the hold of the ship. By this means the seed will be killed, whereas the larger mussels survive the treatment.

C.2. The common shore crab (Carcinides maenas)

This crab feeds preferentially on mussels and can do a lot of damage, but it cannot break open the shells of larger mussels. The most ready victims are mussels which are too weak to close their valves tightly, and thin-shelled mussel seed. Small seed collected from the breakwaters has a thicker shell than the larger yellowish seed gathered from the natural beds in deep water such as the Deep of Renesse. Once the crabs attack a consignment of mussels the odour of mussel flesh transmitted by the tidal currents may lure so many more of them that nothing remains of a whole boat-load scattered on a particular plot. Therefore the planting of vulnerable seed should be carefully timed, taking into consideration the fact that crabs migrate to deeper water in winter. Vulnerability may be due to thin shells and to long delay before the seed is planted. In Zeeland the crabs are usually more troublesome than in the Waddenzee.

C.3. Starfish (Asterias rubens)

Larger starfish are able to open mussels of any size by a continuous steady pull, and they then devour the soft tissues by protruding the stomach through the mouth and digesting externally. Though small starfish can only attack small mussels successfully, they can be so numerous that a whole bed of such mussels may be devoured in due course. The mussel farmer has to watch carefully for the appearance of starfish on his mussels and take action when their number becomes critical. Large starfish can be removed with the special starfish dredge [6] and small starfish can be eradicated with salt, as has been described above.

C.4. Severe winter weather

Cold weather conditions do not easily kill mussels, since they belong to the boreal fauna and can withstand low temperatures very well. If the mussels are living high in the intertidal zone, ice may form on their shells causing them to float away. Drifting floes may destroy large numbers of mussels on natural beds which are not deep enough to escape the moving ice; the mussels are usually carried to sites where survival is impossible. Promising quantities of young mussels on the natural beds and on some plots in shallow water can be destroyed in this way in a severe winter, which occurs in The Netherlands approximately once in every 5 years.

C.5. Fresh water

A drop in salinity does not easily kill the euryhalinic mussel; it can cause the mussel tissues to swell, which wrongly gives the impression that the mussels are of prime quality. On Waddenzee plots which come under the influence of fresh water sluiced out from the IJsselmeer, the stomachs of the mussels are often found to contain freshwater plankton. On such plots growth slows down and in due course the mussels may begin to show an alarming rate of mortality. The experience of the mussel farmers is that transplantation, the usual remedy for any mussel infirmity, does not help very greatly in this instance. Abrupt transition from a low salinity plot to a high salinity plot usually results in large scale mortality. The mussel farmers' experience shows that the death of starfish caused by low salinity can be taken as a warning. In Zeeland low salinity rarely worries the mussel farmers but in the Waddenzee, plots in the vicinity of the sluices of Den Oever and Kornwerderzand become a problem, especially in years with excessive rainfall.

C.6. Gales

Gales can cause much havoc on mussel plots which are insufficiently sheltered. The deeper the plots, the less is the risk of damage. Since the plots in Zeeland are on average 6 to 7 m below the low water level, but only 2 to 3 m in the Waddenzee, damage is more often suffered on the Waddenzee plots. The low production on Mr. Zuijdweg's Waddenzee plots in the season 1966—1967 can be ascribed to gale damage. Wind-force 9 gives cause for serious concern in the Waddenzee, and one can expect gale damage here at least three or four times in a period of 10 years. Gales occurring in summer are less dangerous as a rule, since the mussels are able to withstand them more effectively at higher temperatures; after summer gales the farmer usually observes that the mussels are buried in the beds, but they emerge again in due course. Gales often cause some shifting of the sea bed layers. Mussels can survive when a layer of sand 2 to 3 cm deep covers them, but are smothered when more than 5 cm of sand is deposited on them as a result of gales.

C.7. Silt

Silt can give cause for concern if it is deposited under the mussels in excessively thick layers as a result of their own filtering activity. This can lead to erosion of the mussel carpet, and sometimes also to a reduced supply of food because of the reduced current velocities, so that growth is therefore slower. It has been described above how such plots should be worked with a heavy chain-harrow at a spring tide. It is still questionable whether the high oxygen consumption of biodegradation processes in silt which is rich in organic matter, may lead to reduced filtering rates in the mussels on the plots.

C.8. Barnacles (Balanus crenatus)

These barnacles may settle on the shells of the mussels. Though they live on planktonic food they can hardly be called competitors, since they do not noticeably affect the quality of the flesh of the mussels on which they settle. In this respect the bryozoan *Membranipora*, which also often settles on the shells of the mussels, is no different. The mussel farmer nevertheless considers only the barnacles as a serious mussel pest, for they are difficult to remove from the shells, and hence are found in the cook's pot. Boiling releases small sections of their calcareous shells which are difficult to remove and may spoil the dish. Consignments of mussels with many barnacles on them fetch therefore a lower price at the auction, usually about 25% below normal. A very heavy settling of barnacles is sometimes better than a light settling, for it often happens that the periostracum eventually peels away from the shells when the latter are thickly plastered with barnacles. If a secondary settling takes place however, it will never be released from the denuded shells. There is little the mussel farmers can do to prevent or counteract barnacle settling, but they know from experience where and when settling is most expected.

C.9. Polydora ciliata

This is a polychaete worm, superabundant in many coastal waters, which lives in a burrow dug in any of a variety of materials, molluscan shells included, and this burrow is often extended by a tube built with mud particles. Oysters may suffer severely when many Polydora dig into their shells, especially when the inner surface is reached and must therefore be resealed with conchyolin. Such honeycombed shells are moreover brittle, and often break during transport. This damage rarely occurs in mussels, for *Polydora* cannot burrow through the thick, tough periostracum layer of the mussel shell. Only when the periostracum has come off, for instance after heavy barnacle settling, can *Polydora* enter.

In the period under consideration there was one exceptional year, 1963, in which *Polydora* caused serious damage to the mussels on the plots in the Waddenzee. Early in the summer the mussel farmers found that the mussels were difficult to collect, that there was a dark brown felt-like layer about 1 cm thick, which covered the mussels like a carpet; the mussels were also beginning to die. In the dying mussels they found conspicuous long red worms which they blamed for the disaster. The worms, *Capitella capitata*, were however attracted by the odour of the dead mussels, and were feasting on them. It was *Polydora* which had created the felt-like carpet, smothering the mussels. Though *P. ciliata* is very common, it may have been the large scale turnover of chemical components following the unusually long winter of 1962—1963, in which whole communities of marine organisms were killed off, which led to an extraordinary level of reproduction in *Polydora*. Several farmers lost an appreciable percentage of their stock that year.

C.10. Mytilicola intestinalis

This parasitic copepod, usually called the mussel parasite, had never been observed in the Dutch coastal waters prior to 1949. It was well-known in the Mediterranean and was found on the northern coast of Germany in the late nineteen thirties. In 1948 it was accidentally introduced into the Zeeland mussel area in a boat-load of infested mussels fished from the Dutch—German border area. In Zeeland it proliferated explosively in the densely-populated mussel districts of the Zandkreek and the Oosterschelde. Though even one single *Mytilicola* exerts a measurable effect on the quality of a mussel, serious trouble ensues when, in a period of higher water temperatures, ten or more parasites invade each mussel. Such a disaster occurred on the Zeeland beds in 1950, reducing drastically the total Dutch production of mussels. On the advice of the fisheries biologist in charge of shellfish investigations, the mussel population on the plots in the Zeeland area was reduced to about 30% of the usual density. This did not eradicate *Mytilicola*, but brought the infestation of the mussels down to a level which allowed normal growth and fattening.

In addition, an effort to grow mussels on plots in the western Waddenzee was advised. This effort was initiated in 1950 and led in due course to the complete restoration of the previous level of Dutch mussel production. Though there are fluctuations from year to year, one can say that on average 70% of the Dutch mussels come from plots in the western Waddenzee, which is still free from *Mytilicola.* Since *Mytilicola* does occur nowadays in mussels on the natural beds in the Waddenzee east of the Isle of Ameland, it is strictly forbidden to transfer these mussels to the plots in the western Waddenzee, although one may get permission from the authorities concerned to transfer them to plots in Zeeland.

It seems almost unavoidable that the mussel parasite will make gradual advances in a westerly direction and will one day reach the plots in the western section of the Waddenzee. Serious damage need only be feared in areas where large numbers of mussels are concentrated in a small volume of water with a limited degree of flushing. Under such conditions most of the larvae of the parasite are retained in the mussel area. The prospect is therefore, that mussel farming will remain possible at the present level in the western Waddenzee even when *Mytilicola* makes its appearance there.

C.11. Mussel disease

As in all other cases of farming marine organisms, the greatest threat may come from the smallest organisms, pathogenic microbes or viruses. In the early years of this century, a serious level of mortality occurred on the mussel plots in Zeeland, which reduced the production drastically and threatened to ruin the industry completely. From the evidence available, there is every reason to assume that this disaster, called "mussel disease", was the result of an attack by a virus or microbe. It may be assumed that neglect of the plots, insufficient cleaning and incomplete removal of older mussels, provoked an outbreak of the disease. This may be true, for it is the general experience that certain diseases spread unduly during periods of neglect, overpopulation, or malnutrition, but one should not rely too heavily for their control on the fact that the plots are now kept clean by mechanical means. There was a slight recurrence of the disease in a sector of the Grevelingen in 1937. Since then no mass mortality of mussels through unknown agents has been observed in the Dutch mussel districts.

D. THE ECONOMICS OF MUSSEL FARMING

Mussel farming Dutch-style is a full-time business and requires skilled labour.

D.1. Rents

One can only be a mussel farmer after one has taken a lease on one or more plots. The rent which must be paid for the plots is not a heavy burden on the enterprise, as it does not fully reflect all the various services the government provides for the mussel farmers. The plots in the Waddenzee, nine of which are rented by the firm of Zuijdweg, are leased at a uniform rate of Dfl. 300 per year. The rents for the Zeeland plots differ considerably however, according to the quality. The rents paid by the firm of Zuijdweg vary from Dfl. 200 to 900 per plot per year.

D.2. Inventory items

A well-equipped ship, as already described [1], is the most expensive inventory item for the mussel farmer. A well-equipped mussel ship nowadays costs about Dfl. 450 000. The hull of the ship, if properly maintained, has a life of 40 to 50 years. The engine, which has to work at full speed for prolonged periods, because of the considerable distance between the farming plots in the Waddenzee and the centre for cleansing, storage and dispatch at Yerseke, has, of course, a much shorter life than the hull of the ship. A new engine is therefore an item which appears on the budget about once every 6 years. The mussel farmer must be aware that he will have to buy a new ship long before the hull is at the end of its life. This is because of the rapid evolution in the mechanization of the mussel farming operations which may make it desirable to replace the whole ship by one which is better adapted to the modern methods of working. Just as is the case for fishing boats operating on the high seas, the owner of a mussel ship must follow modern developments and mechanization quite closely, not only for the sake of the economy of his enterprise as such, but also as a means of keeping his highly skilled skippers and crew. Mr. Zuijdweg's firm has two mussel ships, the YE 34 and the YE 73 in full-time use; in principle one operates in the Waddenzee and one in Zeeland. For the maintenance of these two ships, excluding amortization, he must expect an annual expenditure of about Dfl. 50 000. If there is an overload of work, the firm hires extra ships. For collecting mussel seed this is done regularly at a contract price of Dfl. 4 per 100 kg of seed fished. A day's work on the plots in Zeeland will cost Dfl. 400 per day for a ship and a crew of three.

D.3. Materials

In comparison with the inventory items, the expendable items do not make heavy demands on the budget of the mussel farmer. He must expect to replace all the oak seaposts which mark the plots at least once a year. The oak seamarks are from 7 to 13 m long, depending on the local depth, and cost about Dfl. 1/m. They are placed from 50 to 60 m apart. If kept in heavy use, the mussel dredges have a life of only about 2 years. Replacement of a dredge costs about Dfl. 800. For fishing in deep water, the dredge requires about 165 m of steel wire, double galvanised, 10 mm thick, having a life of about 1 year. The ship's engine requires fuel for the fishing operations and the voyages to and from headquarters. One 3-day trip from Yerseke to the Waddenzee and back requires about 1 400 l of diesel oil. In addition, one should budget for minor expenses, for salt to kill starfish, for rubber boots and oilskins, for forks, shovels, and metal baskets.

D.4. Labour

It is of great importance to have a skilled crew on board the mussel ships. Often the mussel farmer himself, the leader of the enterprise, will work on board his own ship, but for a large enterprise operating more than one ship, or in an enterprise in which the mussel farmer is also a mussel dealer, the director will only occasionally sail himself. The firm of Zuijdweg, operating two ships, has therefore six men permanently in service aboard its ships. Wages and social security payments amount to at least Dfl. 50 000 per year for a crew of three. This means that one ship should land mussels to the value of at least Dfl. 100 000 per year, in order to work on a remunerative basis. The number of working hours per week should not deviate too much from that in other firms, and when necessary employees have to be taken by car from Yerseke to the ports on the Waddenzee.

D.5. Profitability

The Dutch Institute for Agricultural Economy has carried out investigations on the profitability of the Dutch mussel industry. The enterprises were put into the categories "small", "medium" and "large" and it was demonstrated that there was a great difference in profitability, owing to the fact that the larger firms were working on a much sounder basis than the smaller enterprises. The larger firms achieved this partly by spreading the risks and using a greater number of plots in various areas, but above all because they were able to make full-time use of employees and equipment. Idling, including time spent on work of lesser urgency, is the greatest enemy of the mussel enterprise. A modern ship is a costly investment and should not stand idle in port. If one has fewer plots than can be worked with one ship, this inevitably means a lower degree of profitability. Renting ships on a day-to-day basis is only advisable in periods of high workload, not under normal conditions. There should be a good balance between the number of ships, the numerical strength of the crew, and the number of plots worked.

D.6. *Economics of mussel growth*

The basic principle of mussel farming is the growth of small seed into large mussels on well-protected plots. It is the growth and fattening of the mussels which should bring the profit. It should be clearly understood that mussel farming is a semi-culture, not a kind of manufacturing process, and that growth and fattening are to a certain degree at the mercy of the whims of nature. If one starts with one ton of small seed one can expect to double or triple the tonnage during the growth from seed to half-grown mussels (measuring 4.5—5.0 cm shell-length), but it is a common experience that one cannot rely too much on an increase in tonnage during growth from half-grown mussels to mussels of marketable size. Losses caused by a variety of factors, listed under the heading "farming risks" usually erode any gain in weight by growth in that period. Occasionally one may have the good fortune to harvest more than the tonnage of half-grown mussels present at the start, but the converse occurs more frequently.

D.7. *Production limitations*

From the foregoing it can be seen that the firm of Zuijdweg lands about 1 600 000 kg of mussels per year, about 50% from the Waddenzee plots and the other 50% from the plots in Zeeland. Of the total annual production in The Netherlands this firm contributes about 2%.

For the time being, the firm of Zuijdweg sees no need for expanding its production. The limiting factor is the area of suitable mussel plots. A significant increase in area brought about by renting two or three additional plots would inevitably create a shortage of ship-time, leading either to insufficient maintenance of the plots or to considerable expense for the hire of extra ships. A much greater increase in the area of growing plots would be required to warrant the necessary investment for another full-time ship. Owing to the risks involved in mussel farming, it is advisable for a firm which is at the same time farming and dealing in mussels, to maintain a certain financial balance between the two sides of the business.

E. MUSSEL TRADE

E.1. *Cleansing and storage*

During operation, the mussel dredge swirls up large clouds of sand and silt. Mussels filtering the water in search of food may be taken by surprise by such a dust cloud and may be unable to prevent some sand entering between their valves. If the dredge picks them up immediately, there will be no opportunity to dispose of this sand and silt. Hence, freshly gathered mussels always include specimens which contain some sand in the shell cavity. Such mussels are unacceptable to the consumer. One should therefore free the mussels carefully of all the sand and silt which may have entered their valves during fishing operations.

For this purpose, the Dutch mussel dealers use cleansing plots. All the cleansing plots now in use are situated in the vicinity of Yerseke, some to the northwest of Yerseke (Geultje) and some to the southwest (Harden Hoek). There are altogether 30 plots measuring 200 × 250 m each, quite high in the intertidal zone and close to the land, with a clayish subsoil so stable that the human foot does not sink into it to any extent. These cleansing plots are very well sheltered from the prevailing southerly and westerly winds by the dikes of the mainland.

Mussels planted on these cleansing plots are covered with sea water most of the time and therefore have ample opportunity to eject any sand or silt that may have entered between their valves. They can, moreover, recover from the rigours of transportation and regain enough strength to be transported live to the markets in Belgium and France.

In this phase of production one should be aware that the mussels are a product which should meet high standards of quality. Not only must they be free of sand and silt, but, since at least part of the mussel crop is eaten raw on the half shell, they should also be free of germs and viruses which may be a danger to man. They should, therefore, always remain in water which is bacteriologically clean. Another important requirement is that there should be no dead mussels in the living crop, for although it may not be dangerous to swallow a mussel which has been dead for some days, the experience may cause the consumer to develop a dislike for this product.

Great care is therefore required. The water covering the plots at high tide should have a high, uniform salinity, and should be agitated constantly by the tide to keep it saturated with oxygen. But the mussels must be distributed on the cleansing plots in such a way that all of them are accessible to the pure water.

As soon as possible after the load of mussels has been sold, the ship is brought to the cleansing plots. The mussel dealers' organisation assigns plots to each member and every dealer has at his disposal the services of a specialized employee called "superintendant of the cleansing plots". This employee guides the ship to an empty section of the cleansing plots and supervises the correct planting of the mussels with the aid of conveyor belts in the hold of the ship. After about 30 min the load is completely discharged on the cleansing plots, for the belts are moved at a higher rate than is the case when mussels are planted on the deeper growing plots. Planting the mussels too thickly should be carefully avoided; about 10 kg/m^2 is considered the correct rate, which means that the entire plot can accommodate about 500 000 kg of marketable mussels. It is impossible to avoid laying some mussels over others, with the possible result of smothering. On a normal growing plot this would

Fig. 23. When a load of marketable mussels is discharged on the cleansing plot at high tide they are carefully spread by hand at low tide to avoid smothering. (Photo A.C. Drinkwaard.)

cause a certain small percentage to be lost, but on a cleansing plot, where the mussels remain for a few days only, the result of overlaying may mean dead mussels decaying in consignments for the market. To avoid such a situation the superintendant goes to the cleansing plots at the next low tide, armed with a fork [7], and spreads the mussels as evenly as possible (Figs 23 and 24). If time allows, he can spread about 125 000 kg of mussels, i.e., about a quarter of the plot, in one turn of the tide. The better the quality of the work during planting, the shorter the time required for the final spreading by hand. In periods of high workload, extra assistants are employed. It is unavoidable that mussels which have been cracked during fishing and some starfish will be found in a consignment. Assistance in eliminating these is provided free of charge: at low tide this comes from the seagulls (Fig. 25), and at high tide from the shore crabs. It will be clear from the foregoing that an important requirement for a Dutch-style cleansing plot is that one can sail there with a loaded ship at high tide yet one must be able to walk on the plot for a couple of hours at low tide. This is only possible where the tidal range is quite large, as it is in the Oosterschelde at Yerseke (range about 3.50 m), but not in the Waddenzee where the tidal range is about 1.50 m.

Fig. 24. Even in mid-winter the work on the cleansing plots goes on, unless ice floes make it impossible to continue shipment of mussels. (Photo A.C. Drinkwaard.)

Fig. 25. The seagulls help the mussel dealers by eliminating, free of charge, starfish and damaged mussels. (Photo A.C. Drinkwaard.)

The biological process of cleansing, the ejection of sand and silt, takes little time. Within 24 h all the mussels will be sufficiently clean. It takes more time before the mussels have gained the stamina for a further journey. Experience has shown that mussels brought in from the Waddenzee should preferably remain from 6 to 8 days on the cleansing plots. Those from the Zeeland plots experience practically no change in salinity and travel much less, so that for these, 3 days will suffice. The cleansing plots also serve as wet storage. The market should be supplied regularly, commensurate with demand, but fishing cannot possibly keep pace with demand, not only because of interruption by week-ends and Church holidays, but also because weather conditions such as gales, fog and snow may make fishing impossible. The supply on the cleansing plot helps to smooth out the unavoidable fluctuations in the landings of marketable mussels, and only when severe winter weather makes fishing impossible for a prolonged period may the dealer be unable to supply the market.

E.2. Fishing on the cleansing plots

When the dealer needs mussels for immediate dispatch he gives instructions to gather the required amount from the cleansing plots. Since the cleansing plots have stable beds, there is little danger of stirring up sand during fishing, but to minimise any risk of this, special precautions are taken by rigging the lines in such a way that the dredge moves along in front of the propeller of the ship [8]. The regulations allow fishing on the cleansing plots during the hours of darkness, but there are only a few hours during which the ship has enough water under the keel to work on the plots. The firm of Zuijdweg uses the ship stationed in Zeeland for this work, but in periods when there is a considerable overload of work, an extra ship may be hired. It takes 1 hour to collect 20 000 kg of mussels from the cleansing plots and only about 10 min to sail between the plot and the firm's landing stage.

E.3. Washing, grading and packing

Ashore the mussels are unloaded with a grab [9] which drops them into a hopper at the beginning of a production line consisting of a breaker, a washer, a blower, a sorter, and a metering apparatus. In the breaker [10] the mussels are declustered by a series of rotating knives, water being administered more or less according to whether the mussels show a lower or a higher degree of clustering. The washer [11] is a revolving apparatus consisting of interspaced iron rings into which large quantities of sea water (60 m^3/h) are injected. The quantity of mussels fed into the breaker and the washer varies from 2 500 to 10 000 kg/h, depending on the degree of clustering and on the percentage of tare. The mussels from the washer are dumped onto a wire-gauze conveyor belt under which a powerful blower [12] is installed to eliminate empty shells

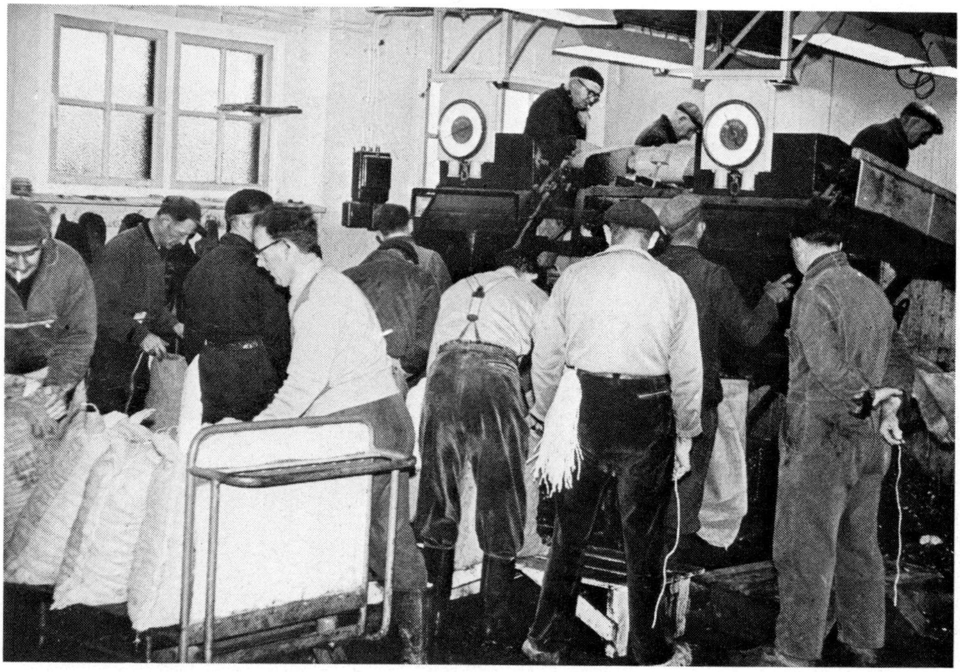

Fig. 26. When both production lines are operating simultaneously a fair number of men are needed to handle the bags. (Photo A.C. Drinkwaard.)

and other light objects; the mussels receive a sprinkling of water to moisten them, and pass on to the sorter [13], a slowly-moving rubber conveyor belt at the side of which four or five persons pick out all undesirable objects which have not been eliminated in the washer or the blower. Finally, the clean mussuls are metered into bags with a special weighing apparatus (Fig. 26) [14]. For export to France burlap bags containing 30 kg are used; these bags are used only once for shipping mussels, although they have already been used for wheat or linseed. For customers in Belgium the mussels are packed in bags of 50, 25 and 12.5 kg, and for some customers, in small plastic bags containing 2 kg. The burlap bags are sewn up by hand with sisal string in such a way that the mussels are tightly packed, thereby reducing loss of water during transport. The bags are then loaded onto a truck (Fig. 27) which will transport them to the customer abroad. The firm of Zuijdweg specializes in export and does not supply the home market.

Fig. 27. The bags filled with clean mussels ready for the market slide down onto the trucks which will take them abroad. (Photo L. Westbroek.)

F. THE ECONOMICS OF MUSSEL TRADING

F.1. Rents

The rent for a plot in the cleansing area (Dfl. 500 per 5 ha plot per year) is not a heavy burden on the mussel trader.

F.2. Personnel

The wages for the highly skilled "superintendant of the cleansing plot" is an item of much greater importance.

F.3. Inventory items

Another important item is the ship on which mussels are collected from the cleansing plot. The firm of Zuijdweg uses the Zeeland stationed ship which is manned with a crew of three, and which also carries out the farming work on the Zeeland plots. Fishing costs are calculated to be about Dfl. 1.50 for 100 kg of mussels.

The production line ashore for declustering, washing, grading and packing the mussels is installed in a building measuring 20 × 12 m. Inside this building the firm of Zuijdweg has installed two production lines placed parallel to one another, so as to be able to keep pace with the demand at the height of the season. The apparatus of the production line handles a considerable amount of sea water, as well as sand, and occasional stones hidden among the mussels, all of which causes so much wear and tear that the lives of the various parts are limited. Moreover, the spaces between the bars of the washing apparatus should be related to the size of the mussels being treated, so that there must be an extra set with a different gauge.

The firm of Zuijdweg employs 12 men on the production line, six of whom are in permanent employment, whereas the other six are seasonal workers. For washing and grading the cost amounts to Dfl. 2 per 100 kg of mussels.

F.4. Expendable items

For packing containers the costs per 100 kg are Dfl. 1.50 when burlap bags are used, and Dfl. 3 for nylon bags.

F.5. Administration

For administration and telephone costs, the expenditure amounts to Dfl. 0.50 per 100 kg of mussels. The firm of Zuijdweg has a small office with one clerk in permanent employment. The book-keeping is put in the hands of a specialist firm at Yerseke.

F.6. Transport

For the transport of mussels the firm of Zuijdweg operates two truck-and-trailer combinations, each of which can carry 30 000 to 35 000 kg. These are housed in a garage of 24 × 12 m. The firm has three drivers in full-time employment. During the height of the season extra trucks are hired from specialist transporters, especially on the peak days Monday, Wednesday and Friday. The trucks are difficult to operate economically, for although there is little or no work for them in the period April—June, the drivers are nevertheless still on the payroll. Transport by road costs about Dfl. 2.50 per 100 kg for journeys such as Yerseke—Liège and Yerseke—Lille, and about Dfl. 4.50 for Yerseke—Paris. Specialist transport is more economic for the longer distances.

F.7. Purchase

The firm of Zuijdweg purchased the following quantities of marketable mussels during the years under consideration, some of which were grown on their own plots:

1963—1964	3 021 700 kg
1964—1965	2 990 900 kg
1965—1966	3 093 800 kg
1966—1967	2 578 500 kg
1967—1968	3 333 000 kg
1968—1969	5 919 700 kg
1969—1970	3 598 300 kg
1970—1971	4 369 500 kg

F.8. Losses

It is very important to realize that there is always a notable discrepancy between the quantities bought and sold. In all phases of treatment between purchase and dispatch, losses in weight occur, partly due to the change in the expression of the weight from gross weight to net weight, and partly due to some loss of mussels, especially on the cleansing plots. According to experience, one should expect a loss from 25 to 28%. When purchased, 1 m^3 of mussels weighs about 700 kg, but eventually only about 500 kg of mussels will leave the country. This explains the discrepancy between the total landings of Dutch mussels from the plots on the Waddenzee and Zeeland grounds (from 90×10^6 to 100×10^6 kg, supplemented by occasional imports from Germany, Denmark and Ireland of about 5×10^6 kg per year), and the quantities dispatched. In the period 1963—1970 The Netherlands exported on average 60×10^6 kg of marketable mussels per year of which France took 35×10^6 kg and Belgium 22×10^6 kg. The firm of Zuijdweg nowadays accounts for about 7% of the total Dutch mussel trade.

F.9. Sales

The price fetched abroad for marketable mussels varies from year to year and differs from city to city, not only as a result of the usual interaction between supply and demand, but also as a consequence of the fierce competition between the Dutch mussel exporters on the French market. In Belgium, the mussels were sold for Dfl. 0.60 to Dfl. 0.70 per kg including transport, in the season 1971—1972, the higher prices being made at the more distant markets, but in France, the mussels were sold wholesale for less than half that price. The retail price varied between BFr. 14 and 28 in Belgium, and between FFr. 2.00 and 2.50 in France, indicating the abnormal discrepancy between wholesale and retail prices in France.

F.10. Seasonal aspect

As has been mentioned before, the sale of mussels follows a marked seasonal pattern, with a close season between a selected date in April and the 1st of July, coinciding with the mussel's spawning season. As soon as the quality of the edible tissues reaches a satisfactory level the sales increase, reaching a plateau between August and November, to decline again in the first months of the year. Occasionally there is some interruption in the sales pattern; about once in 5 years a prolonged period of subzero temperatures may render fishing impossible in January and February. After the thaw, fishing cannot be resumed until the seamarks taken away by the ice have been replaced. In the period under consideration, there was one bloom (in July 1971) of dinoflagellates *(Dinophysis acuta,* and *Prorocentrum redfieldi)* which led to a temporary cessation of sales. The consumption of mussels from the areas where these dinoflagellates were most abundant caused gastro-intestinal disorders. Paralytic shellfish poisoning, following a bloom of dinoflagellates of the genus *Goniaulax* has never arisen from mussels produced in The Netherlands.

G. SERVICES RENDERED BY THE GOVERNMENT

Mussel farming is only possible where suitable grounds are placed at the disposal of the mussel farmers. Therefore, the leasing of plots is the legal basis of mussel farming. Government officials ensure correct delineation of the plots. The rent is nowadays more of a nominal fee than a true rent, and reflects only part of the value of the plot as a site for farming mussels. Formerly the plots in Zeeland were leased by free auction, but nowadays the system parallels that used in agriculture. The old system allowed the lease of a plot for a period of up to 11 years, but nowadays the lease will not be withdrawn from ground which is properly cultivated.

The government protects the mussel farmer against poaching by organizing an efficient police patrol with fast boats. Complaints about the theft of mussels from the plots are rare. Officers aboard ships of the Ministry of Agriculture and Fisheries inspect the mussel areas, collect information on the state of the natural seed beds, assist the farmers in placing seamarks at the corners of the plots after a period of frost, keep an eye on everything that happens on the mussel plots, and are in many respects advisors to the mussel farmers. In addition to this, the Ministry of Agriculture and Fisheries renders a scientific service to the mussel industry. Its Institute for Fishery Investigations carries out research on mussel biology and mussel farming; this work reached a peak in the early nineteen fifties when the mussel parasite *Mytilicola intestinalis* threatened to ruin the industry. Recently, a Mussel Experimental Station has been set up by the government on the Isle of Texel to develop a project for a plant to clean and store marketable mussels after the natural cleansing plots at Yerseke have been closed in the last phase of the Delta-project. The experimental station is a field station of the Netherlands Institute for Fishery Investigations. The Fisheries Institute is also responsible for the frequent analysis of mussel samples, the results of which are used to certify the standard of cleanliness requested by the importing countries. The methods used are based on the assessment of the numbers of thermotolerant lactose-fermenting bacteria after incubation at 44° C, as developed by Clegg and Sherwood, and later adapted by Reynolds and Wood. 5 such bacteria per g of mussel tissue is the acceptable maximum, but as a rule the figures obtained from the analyses show less than 1 thermotolerant lactose-fermenter per g. Apart from this bacteriological examination, which should indicate the virtual absence of domestic sewage in the water where the mussels are grown, or cleansed and stored, the Fisheries Institute frequently studies the phytoplankton in order to predict possible blooms of dinoflagellates. When the data indicate that there is some reason for concern, laboratory tests with white rats are carried out to find out whether the mussels contain a level of the saxitoxine type of toxin, such that the sales of mussels must be interrupted or the opening of the season postponed. All these services rendered by the Government Fisheries Institute are free of charge and are in no way a reflection of the rents paid for the plots.

H. GENERAL IMPRESSION

Mussel farming Dutch-style, in which mussel seed is planted on growing plots and, whenever necessary, thinned out and transferred to deeper plots, has been practised for more than 100 years. Up to 1950 all the plots for growth and fattening were situated in the Zeeland streams, but now from 60 to 70% of the marketable mussels come from the plots in the Waddenzee. The industry has kept pace with technical developments; it has duly taken into account increases in the costs of labour, and is nowadays mechanized to a high degree. In the first part of this century the mussel industry stood in the shadow of the Dutch oyster industry economically, but in the post-war period mussel farming gained in importance and outgrew the oyster industry by developing a high degree of mechanization and by expanding the markets through the use of modern transport facilities. The rapid technical evolution in Dutch mussel farming should be credited to the initiative of the mussel farmers themselves and their close collaboration with local construction firms. Another favourable development was the gradual evolution from a large number of small enterprises to a small number of units of greater size. Profits increase rapidly with the size of the company. A comparison of the Dutch mussel industry with those in Spain and Italy reveals that the Dutch mussels take more time to reach a marketable size. It could be surmized that this difference is due to bed culture as opposed to mid-water farming, or to water poorer in plankton, but there is not the slightest evidence for either assumption.

In all probability the difference is brought about by the fact that the Dutch mussels have to overwinter at temperatures below 10° C for a period of about 5 months, during which the temperature can often drop below 5° C. This decreases the rate of the mussel's filtering activity and results in a serious interruption of growth and fattening.

The length of time allowed between fishing and consumption has decreased noticeably. The rather rough treatment during passage along the production line must be blamed for this. In particular the tearing of the byssus-threads in the breaker gives the mussels such a physiological shock that many of them die, even when returned to sea water. The consumer, however, no longer accepts clustered mussels smothered in silt that require a considerable amount of preparation before they can be put in the pan, but demands a product which is ready for the kitchen, just as vegetables are expected to be nowadays. The demand for small packages and rapid delivery are a logical consequence of this development. Research focussing on this phase of mussel production might lead in due course to methods which cause less physiological stress, but which still meet all the demands of the present-day consumer.

The question arises whether it is necessary for the Dutch mussel dealers to go to the trouble of bringing the mussels to the market alive. Would it not be sufficient to organize the biological cleansing, thus providing a sand-free product, and then to process them in a plant where the mussels are steamed open and shelled automatically, as is already common practice in the Dutch mussel canneries? One could transport the flesh in boxes, using temperature-controlled trucks, and such a procedure would eliminate 75% of the weight, since shells and sea water would be left behind. The edible flesh alone forms an excellent base for some attractive mussel dishes. In The Netherlands some of the mussels already reach the customer in this form, but in France and Belgium, the main markets for mussels, the public wants its mussels alive, still in the shells, just as it likes to buy fish complete with heads and entrails, rather than as fillets. Since "the customer is always right", the mussel dealers have to comply.

A critical time for the Dutch mussel industry will come when the Oosterschelde is cut off in the cadre of the Delta-project (about 1978), resulting in the loss of important grounds for growth and fattening and all the grounds for cleansing and storage in Zeeland. The natural factors governing the productivity of the Waddenzee will probably allow an increase in the production from the plots in the Waddenzee sufficiently great to compensate for the loss of the Zeeland plots, and it is hoped that the results of the investigations at the Mussel Experimental Station on the Isle of Texel will lead the way to the construction of a large plant for the cleansing and storage of all mussels grown in the Waddenzee. Construction of this plant is a duty of the government under the law dealing with the Delta indemnifications, but the exploitation and maintenance costs will have to be borne by the mussel industry *.

* Just now (1975—1976) the Delta-project is being reconsidered, and there seems to be a chance that mussel farming, cleansing and storage can continue in Zeeland.

The present absurd discrepancy between wholesale and retail prices in France, mainly caused by fierce competition among the exporters, leaves room for improvement. It ought to be possible to sell Dutch mussels in France for a more reasonable price without raising the price the consumer has to pay.

I. TECHNICAL DETAILS

I.1. The ship

The modern mussel boat used in Mr. Zuijdweg's enterprise (Fig. 28) is a steel-hulled cutter measuring 25 m length with a beam of 5.20 m. Its keel plates are 10 mm thick, those of the walls 6 to 7 mm. The draught is 1.20 m when the ship is empty, 1.60 m when loaded.

Behind the front deck, which supports a hand-winch for the anchor and a sturdy mast with four booms for hoisting the dredges, comes the spacious hold, measuring 8.00 × 4.60 m, the floor of which is lined with hard tropical Azobé wood 25 mm thick, and the walls with the same wood but half the thickness.

The wheel-house, under which the 250 h.p. engine is installed, is located behind the hold. Inside it one finds the winches which are operated by the wheelman. The crew's quarters are under the after deck.

Fig. 28. Mr Zuijdweg's modern mussel cutter, the YE 73. (Photo courtesy Mr B. Zuijdweg.)

I.2. The dredges

The dredges in use in the Dutch mussel industry consist of a rectangular steel frame, measuring 1.60 m across, to which a bag about 1.10 m long is attached. The underside of the bag consists of iron rings 2 to 3 cm in diameter made of round iron rod 5 mm thick, and the upper side is made of netting with 8 cm meshes, knitted from nylon cord 4 mm thick. The Dutch mussel dredges are not provided with a row of teeth, but fastened to the 38 mm round bar which scrapes along the sea bed, is a "knife", a long rectangular (3 × 1 cm) piece of hard durable steel. Somewhat behind the front bar is the steel or wooden mud-cudgel which extends almost over the whole breadth of the dredge, preventing the dredge from digging too deeply into the mud. Centrally, at the lower end of the bag, is a loop, in those ships in which the emptying of mussels into the hold has been mechanized.

I.3. The winches

The larger mussel boats are equipped with four dredges which can function simultaneously but should be hoisted up individually. Hoisting is done with a winch installed in the wheel-house; this has four drums each of which is equipped with its own handle, and can be operated singly (Fig. 29). The power of the ship's engine is used to operate the winch and the man at the wheel starts and stops the drums. After hoisting up a dredge-load of mussels, he washes away most of the adhering silt by dipping the dredge quickly several times in the sea. Then the dredge is finally hoisted high and swung inboard with the booms.

Formerly, the deck hands grabbed the bar attached to the lower end of the dredge bag and held it whilst the wheelman released the wire so that the contents of the dredge emptied into the hold. This was heavy work. Nowadays the wheel-house in most of the larger ships has an extra winch with two drums to take over this work. Over the whole length of the hold there is quite a high steel wire, to which two pulleys are fastened some distance apart. Through these pulleys run the wires leading to the two-drum winch and at the farther end of each wire is a hook. When a dredge filled with mussels is swung inboard, a deck hand fastens this hook to the loop on the underside of the bag. Then the wheelman allows the dredge to descend using the main winch so that it turns upside-down, emptying its load of mussels into the desired part of the hold.

I.4. The plots

When the site for laying out a new plot has been selected one first of all puts out some provisional oak seamarks so as to get a general impression of the depth and curvature of the gulleys. Next, a permanent corner pole is

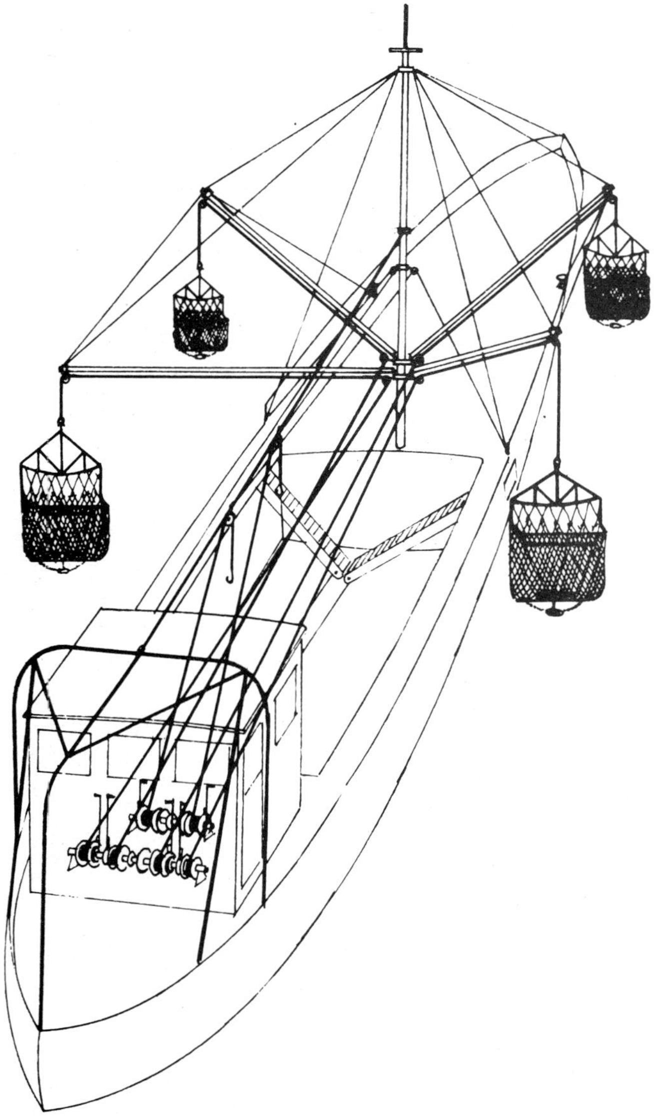

Fig. 29. Scheme of booms, lines, winches and dredges on a modern mussel cutter with mechanical dumping equipment.

placed on the highest part of the ground. For this work one uses creosoted wooden poles, or occasionally iron bars such as old propeller shafts. When the ground is very hard one makes use of a pump unit driven by a 1.5 h.p. petrol engine. With this equipment seamarks and corner poles are installed in a few minutes. The unit consists of a suction/pressure pump, brand Stork KGB, which produces 15 m^3/h of water at 3 000 r.p.m., and to which is attached a flexible tube bearing a long iron nozzle of 22 mm inside diameter. The nozzle is extended upwards with a handle to facilitate its use at depth. The unit is normally operated from a shallop.

To ensure the correct breadth of the plot, one uses (also working in a shallop), a 4 mm thick steel cable of exactly 200 m length, provided with markers at known distances. To ensure that the plot will be exactly rectangular, a reflecting prism or a sextant is used to make angles of exactly 90°.

When ice takes away all the seamarks one has to begin again, starting with the sturdy corner poles, and using land marks where possible.

I.5. Techniques for planting mussels on the plots

A. Chutes. To spread the mussels contained in the hold of a ship evenly on the rearing plots or cleansing sites, the Zeeland mussel farmers long ago developed a special device known as a mussel chute (Fig. 30). These chutes are

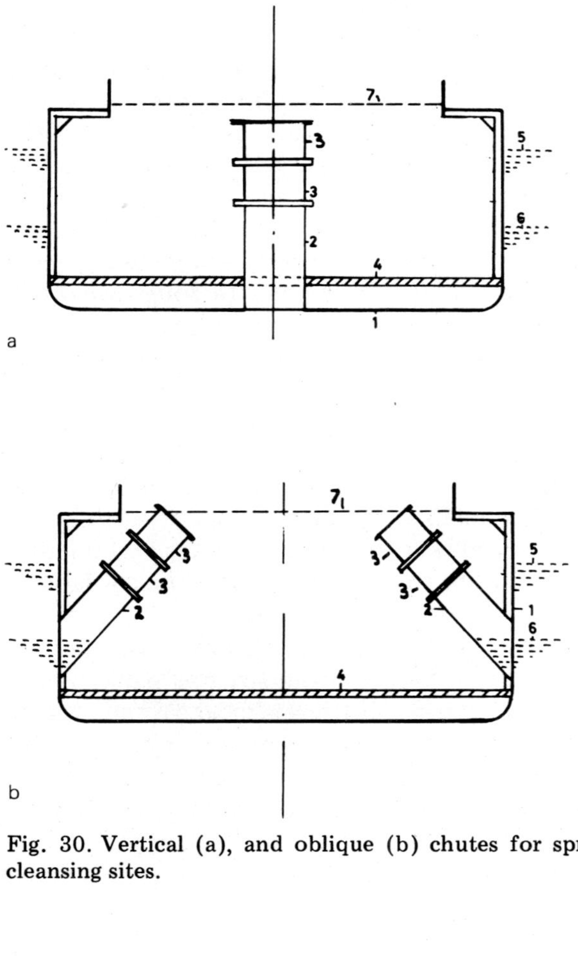

Fig. 30. Vertical (a), and oblique (b) chutes for spreading mussels on rearing plots or cleansing sites.

of two types, (a) vertical, and (b) oblique. If vertical chutes are to be installed, two are placed along the centreline of the hold. They are steel tubes 60 cm in diameter, which open through the ship's bottom (1) and which perforate the wooden lining of the hold (4). They usually consist of three parts: a long lower tube (2) and two shorter removable rings (3) above. The three parts are connected to form a watertight unit by means of four tumble-bolts with butterfly nuts and grease packing. The undermost section (2) is 100 to 120 cm high and for safety reasons should extend to such a height that the upper end is always above water level when the ship is empty (6). The removable sections (3) usually measure 50 cm and 30 cm respectively in height. During fishing operations lids are fixed in place over the ends of the topmost sections. When the hold is filled completely (7) the chute lids are just covered. The waterline of the loaded ship (5) is now higher, but should always remain well below the upper ends of the topmost sections of the chutes. The top of the load of mussels (7) remains 30 to 40 cm below the top of the hatch coaming. If so desired for a longer voyage, the hold can be covered either by wooden hatches put over the coaming, or by a tarpaulin.

To plant the mussels one removes the lids covering the chutes. A mussel fork with blunt prongs is used to shovel the mussels gradually into the chute, and during this operation the wheelman sees that the ship moves slowly and systematically over the plot. When the depth of the load is reduced sufficiently one can take away the top sections of the chutes, and in due course the middel section also. Initially the work is easy as the mussels slowly fall towards the chutes, but as the mussel level falls and becomes lower than the rims of the fixed lower sections of the chutes, the work becomes heavier and heavier for the three or four men working in the hold. In rough weather, moreover, water may come in through the chutes, in which case it is better not to remove the middle sections, and much extra effort is required to throw the mussels from the floor of the hold into the top of the chute.

In order to obtain a more even distribution of the mussels on the plots and the cleansing sites the oblique chute was developed (Fig. 30b). Two such chutes are installed on the starboard side, and two on the port side. The principle of construction and operation remained virtually the same as those of the vertical type, but it was soon realized that the oblique chutes had the disadvantage of taking in more water during rough weather. The heavy labour under such conditions cannot be avoided with these chutes.

B. The screw-jacked false-bottomed hold (Fig. 31). A new invention reduced the amount of work drastically. A false bottom (3) was made in the hold and this could be moved up and down with screw-jacks. When fishing begins one raises the platform (3) to a half-way position, which reduces the height through which the mussels drop when the dredge is emptied. Gradually one lowers the platform until it has reached its lowest position. The hold is filled to the top, level with the deck.

At the site where the mussels are to be planted the mussels are simply

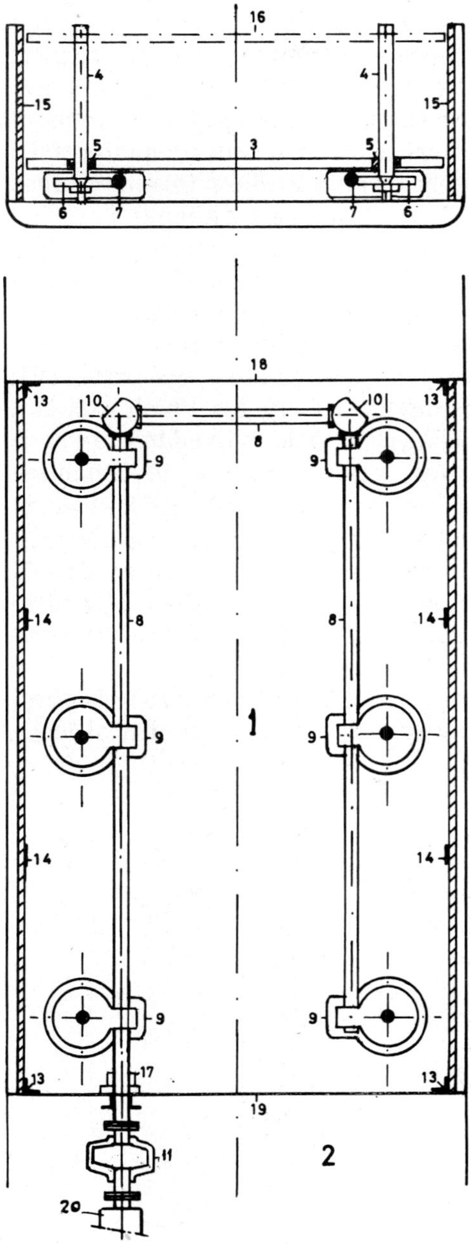

Fig. 31. Screw-jacked false bottom in the hold of a mussel cutter to facilitate shoveling mussels overboard onto the plots.

shovelled overboard with the usual mussel forks. Since there is no bulwark in the way it takes little effort to push them overboard. By raising the platform gradually the mussels remain level with the deck and thus three men can plant a shipload of mussels in 30 to 45 min.

Though this system of unloading offers several advantages over the chutes, especially a reduction in hard work, it was realized in due course that there were certain disadvantages here too. The installation is costly, requires considerable maintenance and repair and suffers from corrosion. It is important that the walls of the hold (15) are precisely vertical to allow free movement of the platform. There must be less than 1 cm clearance between the moving platform and the wooden lining of the hold of the ship, otherwise mussels will drop into the space underneath the platform. Therefore, a lining of hard tropical wood such as Azobé is necessary and flat iron bars (14) are attached to it to guide the platform. For raising and lowering the platform (3), there is an auxiliary diesel engine of 40 to 50 h.p. installed in the engine-room (2). Its shaft (8) has to pass the watertight bulkhead (19), for which reason a packing box (17) with the gland at the engine-room side, is fixed to the watertight bulkhead. The diesel engine (20) drives a reverse gearbox (11), enabling the shaft (8) to turn in either direction. The shaft drives the oil-filled worm gears (9). The worms (7) are connected to the shaft (8), and the worm-wheels (6) drive the screw spindles (4). Large bronze "nuts" (5) which are replaced every year because of corrosion and wear, are fixed to the platform. By turning the threaded spindles the platform moves up or down on the threaded "nuts" (5).

To avoid undue wear it is of the greatest importance to stop the platform automatically when it has reached its lowest position. At the critical level,

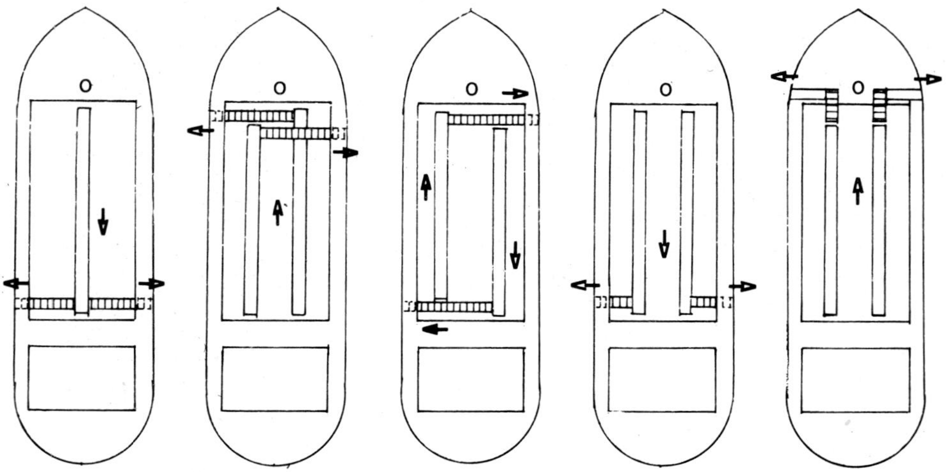

Fig. 32. Different ways of installing conveyor belts for unloading mussels.

therefore, the platform touches a lever connected to a wire. The tension in the wire then operates the stopping mechanism of the diesel engine.

C. Conveyor belt system. In recent years the need to economize still further on heavy manual labour was the spur which led to the development of a new and highly efficient system of planting mussels on the plots (Figs 32 and 33). The principle is quite simple; a system of conveyor belts moves the mussels upwards to wash-ports made high in the wall of the ship so that a gradual stream of mussels drops overboard while the ship moves slowly and systematically over the plots.

The advantage of the system is that both the horizontal longitudinal belt on the floor of the hold and the ascending transverse conveyor belt installed along the short side of the hold are mounted in steel frames, and can be covered with a series of detachable steel lids so that they can move freely even when the hold is fully loaded with mussels. The conveyor belts are either moved by the ship's engine or by an auxiliary engine via a shaft, gear wheels and chains.

When the planting operation begins, the conveyor belts are started and the topmost lid is removed from the ascending belts. The mussels in the immedi-

Fig. 33. Construction of conveyor belts, the lids taken off from the ascending belt, whereas the longitudinal belt is covered. (Photo L. Westbroek.)

ate vicinity drop onto the belt and are carried overboard. Soon a depression appears in the surface of the mussel load and when no further mussels fall on to the conveyor belt the second lid is removed. This continues until one reaches the bottom of the hold where the lifting belt begins. When this position is cleared of mussels, one takes away the first lid of the horizontal conveyor belt which then guides more mussels towards the ascending belt. This is continued until the hold is practically empty. With the fork the last mussels are thrown on to the horizontal belt, and in this way a load of mussels can be planted in a remarkably short time with the minimum of labour.

Various arrangements of the conveyor belts can be tried. Two horizontal belts on the floor of the hold instead of one at the centreline ensure that virtually all the mussels will be picked up by them, eliminating further the amount of work with the fork. Often the ascending belts discharge at the front section of the ship, but some farmers feel that discharging the mussels further aft gives improved spreading on account of the turbulence brought about by the ship's propeller. The diagonal pattern of ascending belts offers the advantage of a more gradual emptying of the hold so that the ship keeps an even keel during the entire operation.

The conveyor belts are made of rubber 1 cm thick and have a breadth of 40 cm. The horizontal belts are smooth, whereas the ascending belts have crosswise rubber strips 3 cm high, about 30 cm apart. The steel lids which are placed 4 or 5 cm above the belts measure 40 × 40 cm and an eye is welded onto them to facilitate their removal using a long-shafted hook. When out of use, the lids are hung on hooks on the walls of the hold. The speed with which the belts move can be adjusted, the ascending belts moving somewhat faster than the horizontal ones.

It is interesting to note that a similar system of conveyor belts with removable plates is used on some oyster barges in Willapa Bay, Washington, U.S.A.

I.6. The starfish net

The principle of the starfish dredge (Fig. 34) is that a thick roller with a roughened surface passes over the mussel beds at a given speed and throws up the starfish that it touches, and possibly also some of the mussels. The heavier mussels sink rapidly again, but the lighter starfish more slowly. If a light net is towed some distance behind the roller it will catch the starfish during their slow descent, but will pass over the mussels. It is of paramount importance to rig the roller and the net in the correct way, according to the speed of towing. The roller, approximately 150 cm long, is of wood and turns freely in the frame holding it. To make its surface still rougher, so that it will adhere to all the starfish over which it passes, strips of old rubber bicycle tyre are often attached to it. Chains of about 60 cm long, fastened to the frame at the same positions as the roller axle, support a light wooden boom, and to the frame itself is welded a horizontal iron bar which stands high above

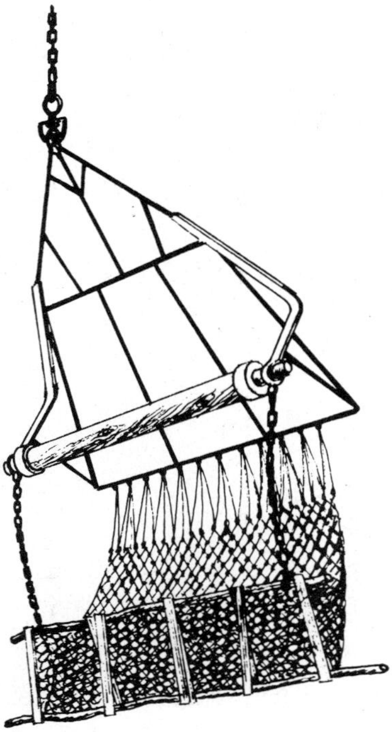

Fig. 34. Starfish dredge with wooden roller.

the roller. To this boom and bar is attached some netting to catch the starfish; this consists of hexagonal chicken wire underneath, and nylon netting for the upper side. At the far end of the net there is a wooden stick protruding at both sides which facilitates the handling and dumping of the starfish in the hold of the ship. Rubber bands connect the front boom to this stick. The entire construction should be so light that it does not dredge up any mussels, but catches only the starfish during their slow descent after having been thrown up by the roller as it passes over them.

I.7. Mussel forks and shovels

Mussel forks bear slightly curved prongs 35 cm long placed quite close together. The ends of the prongs are not sharp but carry blunt knobs to avoid damaging the shells. Usually the forks are seven-pronged, but some farmers use a nine-pronged fork. The blades of the shovels used in the mussel industry are laterally bent upwards.

I.8. Rigging the dredge in front of the propeller

To avoid contamination of the mussels collected from the cleansing plots with stirred-up sand, the dredges are rigged in such a way that they operate

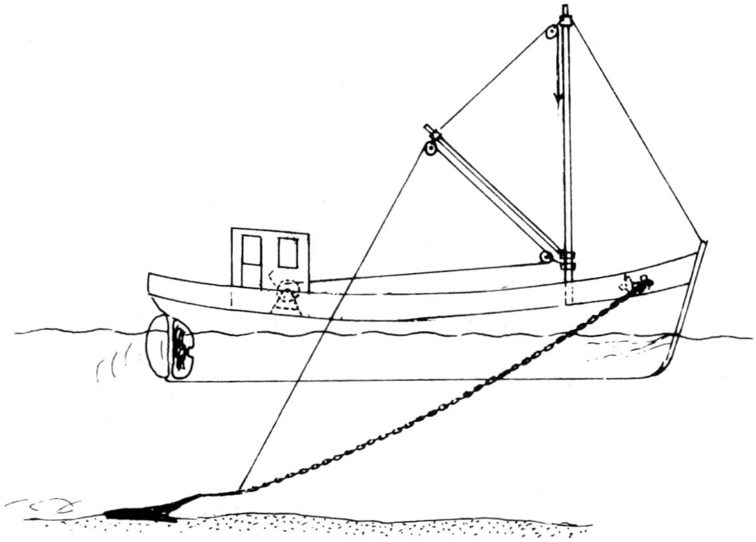

Fig. 35. Rigging the dredge in front of the propeller while fishing on the cleansing plot.

just in front of the propeller (Fig. 35). One thus has to fish with two dredges only, one operating on the starboard side and one to port. The dredges are rigged in the normal way to the wires of the winch passing through the pulleys at the end of the booms, but in addition, chains are fastened to the dredges to connect them to the bollard at the stern of the ship. These chains are from 30 to 35 m long and bear markers at intervals of 1 m. The length of the chains should be carefully adjusted according to the length of the ship and the depth of the water, so that the dredges operate just in front of the propeller.

I.9. The grab

The grab (Fig. 36) used to transfer the mussels from the hold to the production line lifts 300 kg (0.5 m^3) of mussels at a time. It is driven electrically with oil-pressure transmission. An important feature of the grab is that it does not close completely; a chink of 3 or 4 cm remains open so that mussels will not be crushed in the closed jaws.

I.10. The breaker

When the mussels are clustered they should first pass through the breaker, as the present day customer wishes to buy single mussels. The hopper guides the mussels into the breaker, a cylindrical iron tube 1.50 m long with a diameter from 40 to 50 cm. This tube is supported horizontally and inside it turns

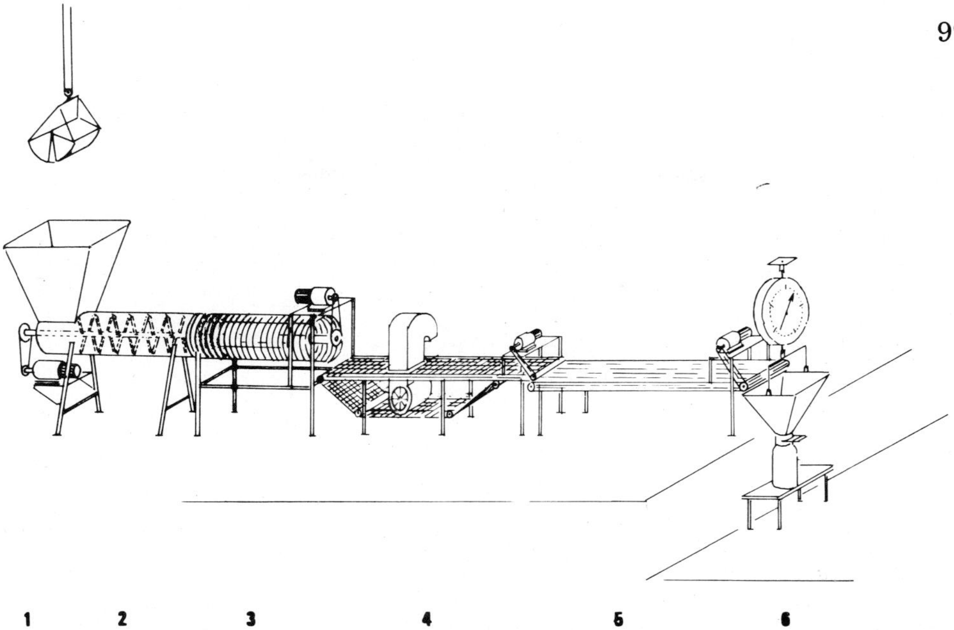

Fig. 36. Scheme of the production line: (1) grab and hopper; (2) breaker; (3) washer; (4) blower; (5) sorter; (6) weighing apparatus.

Fig. 37. Production line in the Mussel Experimental Station on the Isle of Texel. (Photo L. Westbroek.)

Fig. 38. Axis with knives placed in three spiral gangs, taken out of the breaker. In the background, a breaker and washer. (Photo L. Westbroek.)

a hexagonal axle to which sharp coulter-shaped knives made of a very hard steel have been welded in three spiral gangs (Fig. 38). The knives are located at a slight angle and their spiral-wise mounting forces the mussels forward through the tube. Usually the axle turns at 34 r.p.m., driven by a 5.5 h.p. engine. To work efficiently, the tube should be tightly filled with mussels and water. When the mussels are heavily clustered, less water is required than is necessary when they are more lightly clustered.

I.11. The washer

The washer is a rotating cylinder supported at an angle of about 10°, consisting of a series of iron rings. The washer is 2 m long with a diameter of 70 cm. The rings are made of round iron rod of 1 cm diameter and are placed so close together that no mussels can fall through. For small mussels one uses a washer with rings placed closer together than those used for washing larger mussels. A 3 h.p. electric motor continuously rotates the washer at 34 r.p.m. Sea water is forced in through perforated tubes at 60 m^3/h from a pump driven by a 5 h.p. electric motor. The inclined position of the washer causes the mussels to pass through gradually. From 5 000 to 10 000 kg of mussels can pass through the washer per hour if they are already reasonably clean, but only 2 500 kg when a lot of sand, silt and shell debris are present.

I.12. The blower

From the washer the mussels drop on to a conveyor belt of wire netting 4 m long and 60 cm across, moving at a speed of 11.5 m/min. A strong upward current of air directed from under the belt eliminates empty shells, leaves of seaweeds such as *Ulva*, and other light material. This air-flow is generated by an electric air pump with a 5.5 h.p. engine, delivering 1 300 m^3/h of air. The stream of air can be adjusted according to the size of the mussels.

I.13. The sorter

From the blower conveyor belt, the mussels drop on to a rubber conveyor belt 4 to 5 m long and 50 cm wide, which moves quite slowly. This is the sorter and four, sometimes five, men stand alongside it to pick out by hand all the undesirable objects which neither the washer not the blower could eliminate, including occasional stones and damaged mussels.

I.14. The weighing apparatus

The declustered and cleaned mussels leaving the sorter are fed into a hopper attached to an automatic weighing apparatus (brand: van Berkel). This measures a desired quantity of mussels into a sack placed under it. Altogether seven men, four at the sorter, one filling the sacks and two closing the sacks, operate the production line, delivering from 2 000 to 4 000 kg of mussels per h.

J. BIBLIOGRAPHY

Anonymous, 1852. De Zeeuwsche mossel-visscherij. Tijdschr. Staathuishoudk. Stat., 7: 370—372.

Bayne, B.L., 1964. Primary and secondary settlement in *Mytilus edulis* L. (Mollusca). J. Anim. Ecol., 33: 513—523.

Cambier, R., 1952. Mosselvisserij in Zeeland. Tijdschr. Econ. Soc. Geogr., 43: 265—266.

De Blok, J.W. and Geelen, H.J.F.M., 1958. The substratum required for the settling of mussels (*Mytilus edulis* L.). Arch. Neerl. Zool., 13(1) suppl.: 446—460.

Delsman, H.C., 1910. De voortplanting van de mossel. Verslag voor den Staat der Nederlanden, Zeevisscherijen, Schelpdiervisscherij, pp. 75—88.

Fokker, A.J.F., 1901. Rapport Mosselenziekte. Appendix II-2544. Minutes of Board Meetings, Bestuur der Visscherijen op de Zeeuwsche Stroomen, J.M.C. Pot, Tholen, 13 pp.

Fokker, A.J.F., 1902. Rapport Mosselziekte. Appendix II-4595. Minutes of Board Meetings, Bestuur der Visscherijen op de Zeeuwsche Stroomen, J.M.C. Pot, Tholen, 12 pp.

Havinga, B., 1932. Austern- und Muschelkultur. Handbuch der Seefischerei Nordeuropas, 7(5), Schweitzerbart'sche Verlagsbuchhandlung, Stuttgart, 64 pp.

Havinga, B., 1956. Mussel culture in the Dutch Waddensea. Rapp. Proces-Verb. Réunions, Cons. Perm. Int. Explor. Mer, 140: 49—52.

Hildebrandt, A.G.U. and Smit, W., 1963. Bedrijfsresultaten van in Zeeland gevestigde mosselkwekerijen. Aanvoerseizoen 1960/1961 en 1961/1962. Versl. Landb. Econ. Inst., No. 29, 74 pp.

Korringa, P., 1950. De aanval van de parasiet *Mytilicola intestinalis* op de Zeeuwse Mosselcultuur. Visserij Nieuws, 3, suppl., 7 pp.

Korringa, P., 1951. Le *Mytilicola intestinalis* Steuer *(Copepoda parasitica)* menace l'industrie moulière en Zélande. Rev. Trav. Off. Sci. Tech. Pêches Marit., 17(2): 9—13.

Korringa, P., 1951—1955. Epidemiological observations on the mussel parasite *Mytilicola intestinalis* Steuer carried out in The Netherlands. Ann. Biol., 8: 182—185; 9: 219—224; 10: 197—200; 11: 184—186; 12: 230—231.

Korringa, P., 1954. De biologische achtergrond van de recente structuur-veranderingen in onze mosselcultuur. Visserij Nieuws, Suppl., 7: 1—18.

Korringa, P., 1955. Qualitätsbestimmungen an Miesmuscheln und Austern. Arch. Fischereiwiss., 6: 189—193.

Korringa, P., 1956—1958. The quality of the marketable mussels from the Dutch Waddenzee. Ann. Biol., 11: 182—184; 12: 228—230; 13: 241—243.

Korringa, P., 1963. *Polydora ciliata* as a pest in mussel farming. Int. Counc. Explor. Sea C.M., 1963, Shellfish Commun., 80, 1p.

Korringa, P., 1968. On the ecology and distribution of the parasitic Copepod *Mytilicola intestinalis* Steuer. Bijdr. Dierkd., 38: 47—57.

Korringa, P. and Roskam, R.Th., 1960. An unusual case of mussel poisoning. Int. Counc. Explor. Sea C.M., 1961, Shellfish Commun., 49, 2pp.

Maas Geesteranus, R.A., 1942. On the formation of banks of *Mytilus edulis*. Arch. Neerl. Zool., 6: 283—326.

Smit, W. and Davidse, W.P., 1970. Bedrijfsresultaten van in Zeeland gevestigde mosselkwekerijen in 1967/68 and 1968/69. Versl. Landbouw Econ. Inst., Afd. Viss. No. 505, 41 pp.

Tesch, J.J., 1918. Mosselziekte op de Zeeuwsche Stroomen. Jaarversl. Visscherij-inspectie, 24(11) I: 147—151.

Verwey, J., 1951. Het ontstaan van mosselbanken. Versl. Kon. Ned. Akad. Wet., 60(8): 81—85.

Verwey, J., 1952. On the ecology of the distribution of the cockle and mussel in the Dutch Waddensea, their role in sedimentation and the source of their food supply. Arch. Neerl. Zool., 10: 171—239.

Chapter 3

MUSSEL FARMING WITH BOUCHOTS IN THE PERTUIS BRETON, FRANCE

Example
Firm of M. André Bouyé, mussel farmer at Charron, Charente Maritime.
Time of observation: 1971.

A. BACKGROUND

A.1. General principles

The origin of mussel farming on bouchots, a typical example of semi-culture, is said to date back to the legendary Irishman Patrick Walton who was shipwrecked in about 1235 in the Bay of Aiguillon, when transporting a cargo of sheep. He tried to make a living by catching sea birds with nets, and observed that mussels settled profusely on the poles of his contraptions. Ever since in this area, rows of poles have been placed in the intertidal zone in an effort to collect mussel seed and grow it to a marketable size. Despite having a long history, which involves the risk of developmental stagnation, recent years have clearly shown technical evolution in this type of mussel farming. Originally operations were carried out in the very muddy area of the Anse de l'Aiguillon. Rows of poles, called "bouchots", served to catch the mussel seed and in a later phase the seed was transferred to bouchots placed somewhat higher in the intertidal zone. These bouchots consisted of rows of poles interwoven with twigs, so that hedges were created. Those bouchots on which further growth and fattening took place, promoted silting up of the area to such an extent that farming operations had to be moved further and further away in a westerly or southerly direction. For transport in the muddy area originally a kind of "mud-horse" ("acon"), a small light-weight, flat-bottomed boat was used in which one kneeled on one knee, propelled by pushing with the other leg in the mud (Fig. 39). Such mud-horses are used on intertidal grounds in several parts of the world, in areas where soft mud prevails. Nowadays the old "acons" are no longer used this way, though very light boats bearing the same name are still used. Now, both workers and materials are transported in motorboats and it is therefore possible to construct bouchots further and further away from the Bay of Aiguillon. The interwoven bouchots are obsolete now and are even officially forbidden, as they cause rapid silting up and encourage *Mytilicola* damage. Even seed-collecting with poles, practised for ages, has already been replaced to a large extent by the collection of seed on coco-fibre ropes.

Fig. 39. Representation of mussel farming in the Pertuis Breton in the nineteenth century, showing "bouchots" with interwoven branches ("clayonnage") and the mussel farmer in his "acon". (From Fraiche, 1865.)

Planting the poles has been mechanized to a certain extent and long tubes of nylon netting are now used to attach young mussels to the poles of the bouchots where they will grow to marketable size. Despite all the recent improvements, the whole process of mussel farming with bouchots is still laborious. It is the excellent flavour of the mussels from this region which renders mussel farming still possible, despite competition with the larger production centres in The Netherlands and Spain.

A.2. Biology of the species

Little need be added to the information on the biology of the edible mussel, *Mytilus edulis* L., presented in the chapters dealing with mussel farming in The Netherlands and Spain. It is the tendency of the mussel larvae to settle on filamentous material which brought farming with bouchots into existence. The tiny mussels do not settle directly on the bark of the poles of the bouchots, but rather on hydroids which select these poles as a substratum. Only in a later phase do the mussels attach their byssus threads directly to the poles. The poles should therefore be placed in the water several months before the new generation of mussel seed is expected to settle; this allows time for hydroids or other filamentous organisms to settle profusely on the poles.

A.3. Geographical situation

Bouchot farming originated in the Anse de l'Aiguillon, situated on the Atlantic coast of France where the departments Vendée and Charente Inférieure

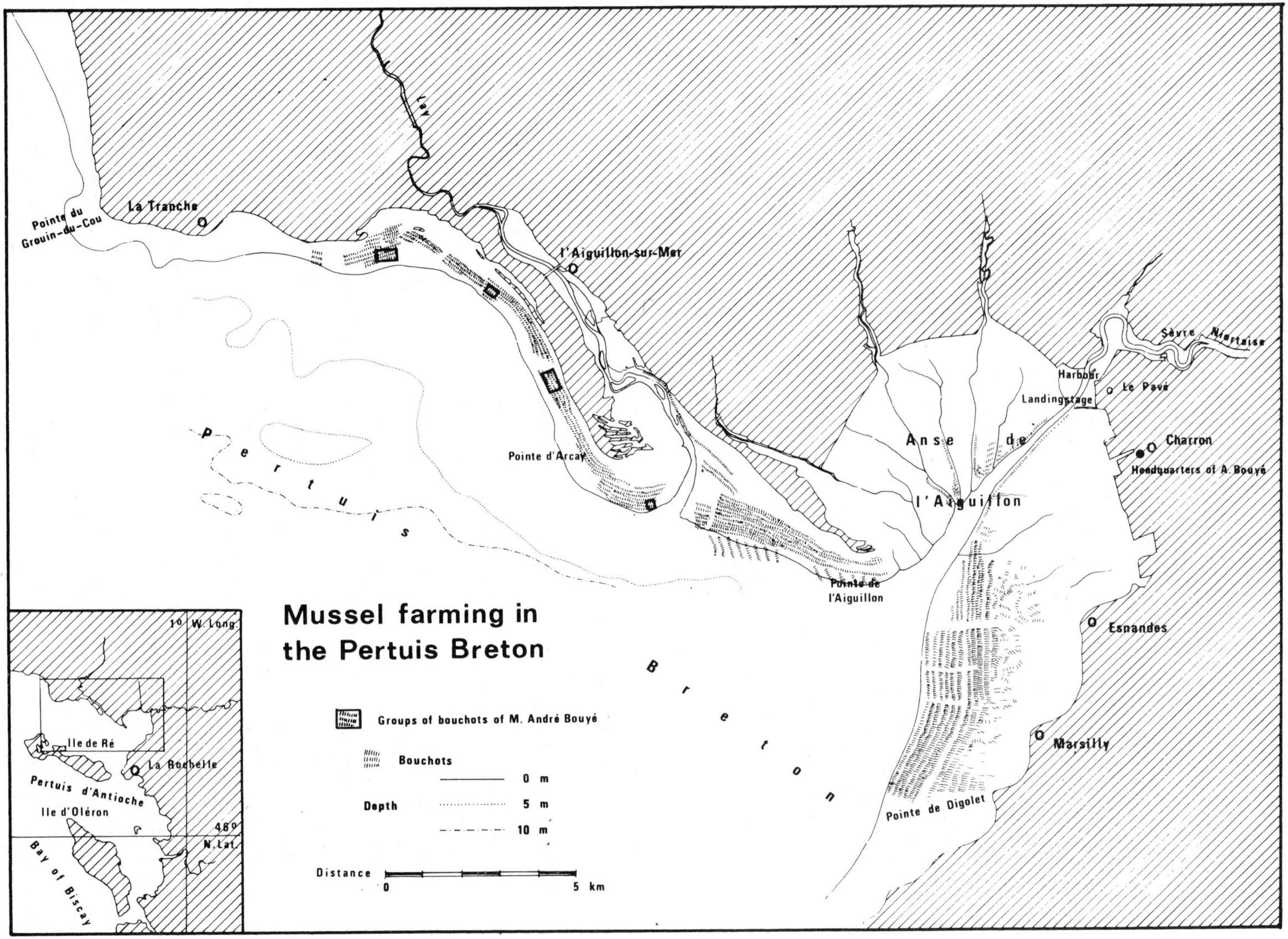

Mussel farming in
the Pertuis Breton
Groups of bouchots of M. André Bouyé
Bouchots
Depth
0 m
5 m
10 m
Distance
0
5 km
Pointe du
Grouin-du-Cou
La Tranche
Lay
l'Aiguillon-sur-Mer
Pointe d'Arcay
Pertuis
Breton
Pointe de
l'Aiguillon
Anse de
l'Aiguillon
Harbour
Landingstage
Le Pavé
Sèvre Niortaise
Charron
Headquarters of A. Bouyé
Esnandes
Marsilly
Pointe de Digolet
1° W. Long.
46°
N. Lat.
Ile de Ré
La Rochelle
Pertuis d'Antioche
Ile d'Oléron
Bay of Biscay

(Charente Maritime) meet (see Chart). It gradually spread over a considerable length of the continental coastline of the Pertuis Breton, the 10 km wide body of water between the mainland and the Ile de Ré, now delineated between the Pointe du Grouin-du-Cou (46° 20′ 40″ N and 1° 28′ 10″ W) and the Pointe de Digolet (46° 13′ 30″ N and 1° 11′ 13″ W). The western-most section of the area consists of sand dunes fringed with a sandy beach extending down to the spring tide low water level. Two narrow sandy spits, ending in the Pointe d'Arçay and the Pointe de l'Aiguillon respectively, enclose the muddy estuary of the river Lay. The horse-shoe shaped Anse de l'Aiguillon consists of extensive mud flats traversed by several shallow gullies discharging the run-off from canals into the bay. From NE to SW a deeper channel, the prolongation of the river Sèvre Niortaise, divides it in two. Further south, calcareous cliffs border the area.

The administrative headquarters of the company under consideration are situated in the village of Charron at 46° 17′ 30″ N and 1° 06′ 13″ W, 15 km north of the city of La Rochelle. M. Bouyé's boat is moored in the little harbour of Le Pavé at the mouth of the river Sèvre Niortaise, 2.5 km NW of Charron. M. Bouyé works almost exclusively on concessions on the north coast of the Pertuis Breton.

A.4. Hydrographic pattern

Depth. The Pertuis Breton, open to the Atlantic Ocean, is generally not very deep. In the central-western part there is a somewhat deeper channel, but along the continental coast the 10 m depth contour is to be found from 4 to 6 km offshore. Mussel farming is carried out on the gently sloping intertidal grounds, a zone over 1 km broad.

Tides. The tidal range varies from about 5 m at spring tides to 2 m at neap tides. Most of the tidal water enters the area from the WNW with maximum current velocities of 2 to 3 knots at spring tide. On the ebb tide part of the water finds its way south, passing through the rather narrow strait between the mainland and the Ile de Ré towards the Pertuis d'Antioche.

Water temperature. The Anse de l'Aiguillon and other inshore sections of the area show a greater seasonal variation in water temperature than the western section of the Pertuis Breton, since air temperatures exert a considerable influence where the water is shallow. The annual temperature range is from about 8°C to 20°C in the western-most section, while in the Anse de l'Aiguillon itself, water temperatures may drop to 4°C in winter, and can attain 24°C in mid-summer.

Salinity. The salinity fluctuates little in the western section of the area, ranging from 32 to 35‰, but further inshore, especially where the rivers Lay and Sèvre Niortaise discharge fresh water, one may observe quite low salinities in periods of heavy rainfall, which may occur in winter and early spring. Salinities as low as 15 or 16‰ may be measured on these occasions, and still lower figures are found near the river mouth.

Plankton and turbidity. No figures have been published on the plankton content of the waters of the Pertuis Breton, but considering the excellent quality of the mussels farmed in the area, sufficient quantities of high quality phytoplankton must be available for months in succession.

A remarkable feature of the water in the area is its notable turbidity. The water bathing the bouchots has a yellowish-grey colour so that a submerged Secchi disc soon disappears from sight; the turbidity increases inside the Anse de l'Aiguillon. This turbidity is caused by finely-dispersed silt, kept perpetually in motion by the tidal currents. Though no exact data on the quantities of sediment per volume of water are available, one may be quite sure that the figures would be considerably higher than those given for the area of mussel farming in the Dutch Waddenzee, and very much higher than those for the mussel regions in Spain and Italy, discussed in other chapters. Presumably the water would be found to contain at least 100 mg/l of particulate matter.

If one observes how "thick" the water is in this area, it is surprising that the mussels can grow and fatten there at all, for mussels must collect their food by filtering edible particles from the water in which they are bathed, and a large amount of inert material must impose a physiological burden upon them. No doubt a lot of this very fine inert material will be ingested by the mussels, resulting in an extra workload for the intestinal tract. This is perhaps the reason why growth is not very rapid and why the mussels marketed are of considerably smaller size than in The Netherlands and very much smaller compared with the mussels from Galicia, but it remains a remarkable fact that the quality and flavour of the mussels found here are excellent, despite the turbid water.

Sea bed conditions. There is a wide range of particle sizes, from sand to very fine silt, in the sea bed deposits in the intertidal zone where the bouchots are placed. Soft deposits are found in sheltered places, e.g., in the lee of the spits and inside the Anse de l'Aiguillon. Waves and currents are the main factors determining the nature of the substratum. That soft mud accumulates so rapidly inside the Anse de l'Aiguillon cannot be ascribed to material brought down by the river Sèvre Niortaise in short periods of rainfall: it is assumed rather that waves and currents re-sort the deposits laid down in the Pertuis Breton, these deposits originating largely from the Gironde estuary but partly also from erosion of the local calcareous cliffs, and that the sheltered Anse de l'Aiguillon acts as a sort of silt trap. Without any doubt the filtering activity of mussels and other kinds of molluscan shellfish which abound here, greatly promotes sedimentation.

A.5. Legal aspects

M. André Bouyé's mussel farm can be described as a private enterprise. He does the work at sea with one full-time hired hand. Occasionally, in times of high workload, another mussel farmer lends a helping hand. Mme Bouyé takes care of the administration and keeps in contact with her husband when he is at sea by radio-telephone.

The intertidal grounds belong to the French State and are administered by the "Domaine Public Maritime" which gives concessions for farming mussels in a given area. As in oyster farming, such concessions are given for a period of 25 years, provided that one uses the ground properly for the purpose described in the deed. Unlike oyster farming, concessions for mussel farming with bouchots are not given in hectares, but in linear units, in terms of metres of bouchot length.

In the matter of concessions, "inscrits maritimes" have preferential rights. M. Bouyé has been an "inscrit maritime" since he was 14 years of age. He takes care of 3 km of bouchots, and also several concessions of 50 m length only. In the section west of the Lay estuary he operates on four groups of about ten bouchots each, but east of the river Lay his concessions are largely split up into single bouchots. This is unpractical and impedes efficient operation. The reason for the single bouchots is the same as that for small units of arable land, i.e., the inheritance of a share of their father's concessions by children who practise mussel farming by profession. Although re-allotment has been considered, it has not yet been implemented.

Some other concessions of M.Bouyé are to be found in the bouchot section south of Charron, but as part of his enterprise, they are of little importance.

The rents to be paid for the concessions, according to the new scale coming into force in 1974, are as follows: for a unit of 250 m and over, FFr. 0.20/m, which means, for instance, FFr. 50 for a concession of 250 m. For every concession under 250 m one has to pay the minimum rent of FFr. 50 per year, so that the smaller concessions starting with units of 50 m are relatively expensive. Officially, there should be a distance of 25 m between two bouchots but often this rule is ignored.

In addition to the rent paid to the owner of the ground, the mussel farmers in this area pay on a voluntary basis a "taxe professionelle" which is levied by the mussel farmers syndicate and legalized by the "Marine Marchande". The level of this tax can be readjusted from year to year but now amounts to 25% of the rent paid per metre of concession. This money is reserved for general purposes such as placing wide-meshed nets at the seaward side of the concessions when invasions of rays are expected, or operating a motor pump to keep the landing stage clean.

Official surveillance is carried out by the "gendarmes maritimes", who, as they have to patrol along the coastline of the entire department, can visit the mussel concessions only very occasionally. Therefore five of the mussel farmers operating in this area act as sworn-in vigilantes, each covering a given section, but their presence does not completely prevent the theft of mussels. It would be more efficient to appoint paid watchmen, but the mussel farmers are very reluctant to accept a substantial increase in the "taxe professionelle".

B. SEQUENCE OF OPERATIONS

B.1. Constructing a bouchot

Bouchots are rows of poles, 50 m long, placed perpendicular to the coastline in the intertidal zone. Sometimes a bouchot consists of a single row, the poles usually placed at intervals of about 20 cm, with about 125 poles per bouchot; otherwise one makes double-rowed bouchots, the rows being 1 m apart, in which case the poles are placed at intervals of about 32 cm, with about 190 poles per bouchot. Though the bouchots should be placed 25 m apart, often there are only 15 to 20 m between them.

The intertidal zone is quite wide here, in some sections over 1 km. This makes it possible to place the bouchots in series end to end. In the deepest part of the intertidal zone, closest to the sea, the bouchots for procuring seed are placed, sometimes in a series of four to six bouchots with interspaces of 25 m. These take up a zone of 300 to 400 m of the intertidal terrain.

On the landward side of the seed bouchots is quite a wide passage, then the bouchots for growth and fattening of the mussels, likewise placed in series of four or five end to end. For growth and fattening one therefore uses shallower grounds than for catching the seed. This is a remarkable feature, for in other countries, in The Netherlands for example, the half-grown mussels are always transferred to deeper beds with stronger currents for growth and fattening. Repeated exposure, on the other hand, will make the mussels stronger, so that they withstand the rigours of transport better than those grown in deeper sites.

Formerly the bouchots for growing and fattening consisted of poles which were placed at intervals of 70 cm and were interwoven with sweet-chestnut branches, the so-called "clayonnage" system, forming complete hedges in which the mussels were placed after having been harvested from the seed poles. Now this system is obsolete and even forbidden in this area. Nowadays these bouchots still consist of single or double rows of poles, usually interspaced by distances of 25 to 30 m, but they are devoid of interwoven branches.

To build a bouchot one needs wooden poles 4, 5 or 6 m long, according to the depth of the site. The lower 3 m of the pole is inserted into the sea bed, so that from 2 to 3 m emerge. Formerly poles of local pine wood were used, but nowadays oak poles, complete with their bark, are used by most of the mussel farmers in this area. M. Bouyé uses exclusively oak poles measuring 12 to 20 cm in diameter at the wider end, and about 7 cm at the opposite extremity. He brings the poles to the building site by boat [1], taking 60 to 80 at a time. It takes two men to insert the poles into the beds. The boat is moored alongside the bouchot (which may be under construction or in need of repair) on the leeside. Carried on board is a 3 h.p. gasoline pump, to which is at-

Fig. 40. Inserting an oak pole into the bottom with the aid of a pump and hose. (Photo P. Korringa.)

tached a 13 m long rubber hose of 5 cm diameter. At the free end of the hose is attached a 3 m long iron pipe of about 4 cm diameter, extended at the union end by a thin boom of several metres length. The pump is set in motion and sea water is ejected from the iron pipe with considerable force. One man places the pipe over the desired position on the sea bed. The pipe can easily be moved downwards, as it blows away the sediments. Now the second man takes a pole over his right shoulder, and carefully inserts its narrower end into the hole made with the iron pipe (Fig. 40). Stooping down he places the pole vertically while his fellow worker ensures that the iron pipe descends further and further alongside the pole. When the pipe itself is out of reach the extension boom serves as a handle. Thus the poles are easily sunk into the bed. When the desired depth has been reached the iron pipe is withdrawn. A few blows with a maul or sledge hammer help to secure the pole in position.

It is clear that this work can only be done when the ground is covered by a fair depth of water, the more water, the easier the work. Two men can place from 60 to 80 poles per day under favourable weather and tide conditions.

B.2. Procuring mussel seed with bouchots

Placing poles for catching mussel seed is invariably done in the period January—April. One expects settling of mussel seed in May and June, sometimes even as late as July, but experience has shown that newly-planted poles will catch no seed. The point is that mussel larvae have a strong tendency to settle on filamentous objects. Among the available objects of this nature, are species of red algae, and in particular, hydroids of various description. Such organisms settle here in the early spring, but are often preceded by a settlement of barnacles. The mussel farmer knows that hydroids may settle profusely on the calcareous shells of acorn barnacles, settlement of which should therefore be welcomed. "Hair" on the barnacles is a good sign. On the other hand, farmers dislike the sharp barnacles when harvesting mussels by hand, especially when they settle later on the mussels. Among the hydroids settling on the poles one finds *Tubularia mytiliflora* as the predominant species. As its name suggests, this species is excellent as a substratum for mussel seed. The mussel farmer, anxious to know whether he will have a good settlement of seed on his poles, takes a tuft of *Tubularia* between his teeth and can then detect whether the mussel seed has settled, even before he can see it with the naked eye. Usually the settlement of mussel seed begins on the deeper parts of the poles about the middle of May. In the western-most section in the vicinity of La Tranche, settlement begins early, and the lower turbidity of the water there promotes the growth of the tiny mussels. The seed is left on the poles until it has reached a shell-length of at least 20 mm.

B.3. Procuring mussel seed with coco-fibre ropes

Since about 1960 a new system for procuring mussel seed has been developed, with considerable success. Since this system has several advantages over the traditional one, it is rapidly growing in importance. The method involves placing two rows of 20 poles about 1.50 m apart, with interspaces between the poles of about 2.40 m in each row. The rows are built in the deeper part of the intertidal zone, again at right angles to the coastline. About 60 cm above the sea bed one nails a thin horizontal pole (about 6 cm diameter) from row to row, and a second one is nailed 35 cm above the first. Over these horizontal poles one stretches 25 interspaced coco-fibre cords, 1 cm thick,

Fig. 41. Two layers of coco-fibre ropes, stretched horizontally, densely settled with mussel seed. (Photo P. Korringa.)

over the whole length (50 m) of the concession (Fig. 41). Thus one obtains a double layer of 25 ropes with a total length for the 50 m concession of 25 km. The ropes can be brought out some time after installing the poles, since the fibres of the ropes serve to some extent as a support for the mussel larvae. The ropes are usually put in place from mid-February to mid-April.

To promote a gradual improvement in the production of mussel seed the "Marine Marchande" has fixed, on the advice of the Fisheries Institute, a maximum for the number of ropes per rack together with a maximum for the number of rope-racks per kilometre of concession (10% of bouchot length). M. André Bouyé now works with 100 m of rope-rack out of his total of 3 km of bouchot concession.

The ropes, which provide a hairy substratum, are said to attract mussel seed somewhat earlier than the vertical poles. The seed grows rapidly on the ropes, especially the lower part which is exposed to air for a shorter time than the upper part. Frequent inspection is a necessity, for if the ropes loaded with mussel seed become too heavy, they can easily break.

B.4. Harvesting mussel seed from the bouchots

The mussel seed should not be left too long on the bouchots on which it settled initially. When the young mussels grow they will in due course leave the original filamentous support, and attach themselves with byssus threads to the poles. When they have settled densely they will sooner or later find it impossible to arrange themselves in one single layer, allowing every mussel access to the water currents from which they collect food and oxygen. Overcrowding may lead to ousting, or the detachment of whole clusters of mussels which then become lost in the soft mud or the shifting sand of the sea bed. The mussel farmer therefore inspects his bouchots frequently to decide when to remove the mussel seed.

Harvesting the mussel seed (Fig. 42) usually begins in the month of August and continues until December, depending on the size of the seed and the density of the population. The mussels are scraped from the poles with a special tool [2] consisting of a sharp curved steel blade attached to a net of metal wire which catches the scraped-off mussels. A long wooden shaft makes it possible to reach all parts of the poles irrespective of the depth of the water. Usually the mussel farmer will work from his "yol" [3], an auxiliary boat about 6 m long, in which he can stand upright throughout.

Fig. 42. Harvesting mussel seed from the bouchot. (Photo P. Korringa.)

Towards the end of the year, when all the seed has been harvested, the mussel farmer returns to clean the poles with a scraper [4], a curved steel blade attached to a long wooden shaft, scraping off barnacles and other epibionts. An entirely new flora and fauna are thus allowed to develop on the poles in early spring, offering the seed of the new generation of mussels a substratum on which to settle. The poles will last for 5 or 6 years, provided that shipworms are not too active.

B.5. Transfer of the mussel seed to higher bouchots

The seed harvested from the bouchots should be transferred to other bouchots reserved for growth and fattening, erected somewhat higher in the

Fig. 43. Making a "boudin" by putting mussel seed in cylindrical nylon netting. (Photo P. Korringa.)

intertidal zone. In former days, one packed the seed into pockets of netting which were squeezed in between the branches of the "clayonnage". Now that "clayonnage" is forbidden, all mussel farmers use the new system which has been in use to some extent for 20 years. With a simple funnel placed on a stand cylindrical nylon nets [5] are filled with mussels, producing "boudins" (blood sausages) of 5 m length with a diameter of about 10 cm (Fig. 43). The mussels fall into these nets in irregular positions. Very soon they will try to rearrange their positions so that their siphons have free access to the open water. Eventually they will crawl through the meshes of the netting, but will remain attached by byssus threads to the shells of other mussels or to the nylon netting, which will gradually disappear from sight (Fig. 44). In order to be able to do this they have to be bathed in sea water. As soon as possible after filling the nets with mussels, the farmer will therefore transfer the nets to the poles of the bouchots which are located somewhat higher in the intertidal zone. Two nets will be attached to each pole, one at the top end, the other somewhat lower. The tide determines which one is attached first. Our mussel farmer uses 7 cm nails to attach the nets to the poles. To attach the upper net he begins to drive a nail into the pole, impaling the top end of the nylon net,

Fig. 44. Since the mussels crawl through the meshes of netting, the nylon soon disappears from sight, but can still be found in the centre of the string. (Photo P. Korringa.)

Fig. 45. Typical aspect of bouchots with larger mussels, still showing the spiral attachment of the "boudins". (Photo P. Korringa.)

which is then wound in a spiral round the pole before the far end is fixed with another nail. The lower net is attached spirally below the upper one.

The mussels continue to grow and gradually crawl out of the netting. For a long time after, the spiral shape of the mussel population on the poles of these bouchots can be seen (Fig. 45). The mussels remain on the poles until they have reached marketable size.

B.6. Transferring ropes laden with seed to bouchot poles

Since the seed on the ropes grows rapidly, the transfer to the bouchots should be carried out in good time, beginning about the second half of July.

It is a very simple operation to transfer the seed attached to the coco-fibre ropes to the bouchots reserved for growth and fattening. One simply cuts the rope into sections of 2.5 to 3 m length, winds these in a spiral round the poles and attaches them with a nail. These mussels are free to move about and will sooner or later attach themselves with byssus threads to the poles themselves. On account of the great number of mussels per metre of rope one attaches only one piece of rope to each pole. The rope itself will gradually decay. Often when the ropes are attached to poles from which marketable mussels have been harvested, some of the latter may remain.

The ropes are practical for the transfer of seed to other areas. One can simply coil them up in little baskets for transport by van. M. André Bouyé has 1.2 km of bouchot concession in the Baie de St. Brieuc in Northern Brittany. Occasionally he transfers ropes laden with seed to these bouchots in his own van. 2.5 km of rope is sufficient to load 1 000 poles with mussels. Direct settling of seed on the poles is often so prolific in the Baie de St. Brieuc, that transporting seed there is unnecessary.

B.7. Harvesting and grading of marketable mussels

The season for harvesting marketable mussels begins here on the 1st of May, and from then onwards mussels are delivered to the market until the end of the season on the 1st of February. The minimum size for marketable mussels was formerly 5 cm shell-length, but nowadays it is only 4 cm. As a rule however, one begins harvesting from those poles on which some mussels were left the previous season, and these mussels will, therefore, have reached a shell-length of 5 cm and over. Next one begins to harvest the mussels which were put on the poles, packed in nylon nets, the previous summer. When bouchots with interwoven branches were still in use, and an over-dense concentration of mussels led to a serious reduction in current velocities, together with overcrowding and malnutrition, the mussels often had to be kept for 2 or 3 years on the growth and fattening bouchots before they could be marketed. For various reasons one now aims at a faster turnover, a sound principle from both the biological and economic points of view.

Harvesting is done by hand from the small "acon" [6] in which the worker kneels down. At low tide he goes alongside a bouchot carrying marketable mussels and grabs, his hands protected with rubber gloves, the clusters of mussels, which are easily detachable. Since the largest mussels are in the outermost positions, the farmer does not aim to remove all the mussels from the pole, but leaves some for a second harvest. When the little acon is fully loaded with mussels the worker proceeds to the larger boat, into which he unloads his mussels with a special shovel [7]. On the starboard side of the larger boat there is a trough [8] for washing and grading the mussels. A good jet of water is directed at the mussels in the trough with a powered pump, and the worker, his hands protected with rubber gloves, rubs the mussels over the trellis bottem. In this way the mussels are washed, freed of barnacles, and separated from one another. Those which are too small fall through the trellis and may be used to make new "sausages" which are brought back to the bouchots.

Some mussel farmers in the area have mechanized this part of the operation by installing an obliquely-placed rotating trellis cylinder equipped with water injection, comparable with the "mussel mill" used in The Netherlands for washing and grading. M André Bouyé is of the opinion that such an investment would hardly be economic, since he always manages to clean by hand all the mussels harvested on a given day, before he reaches his home port.

B.8. Storage of marketable mussels

The mussel farmers of this area are also mussel dealers. They have to supply the market according to its needs, irrespective of weather and tides. For this purpose the radio-telephone installed on the boat is extremely useful, since Mme Bouyé can contact her husband in the field to inform him about orders which have just come in. Though M. Bouyé goes to his bouchots every day, including Sunday, during the mussel season, exceptional weather conditions and unfavourable tides sometimes make it difficult or even impossible to harvest the quantities required. The mussel farmers of the area have no in-

Fig. 46. The mussels, duly washed and graded on board, are packed in woven plastic sacks. (Photo P. Korringa.)

stallations such as storage basins ashore, but they have built some storage cages [9] in the vicinity of their home port. M. Bouyé has two of such "reservoirs" at his disposal, one consisting of six, the other of ten cages. One is close to the landing stage at Le Pavé, and the other is somewhat further away at a lower level. From these cages he can always take some mussels, even under the most unfavourable conditions. Experience has shown however, that such mussels, separated from one another on a previous occasion and thereby physiologically "insulted", are less strong and viable than those harvested the same day from the bouchots. Moreover, the mussels from the reservoirs close to the landing stage are invariably of poorer quality than those from the deeper cages, which are bathed in water of better quality.

B.9. Marketing mussels

The marketable mussels, after being harvested, washed and graded on board, are either packed in rectangular baskets of perforated plastic [10] or in sacks of burlap or of narrow interwoven strips of nylon (Fig. 46), which can hold 25 kg of mussels. Sacks which can "breathe" keep the mussels in better condition during transport than do those made of tight plastic sheets. A simple weighing machine on board the boat is used to fill the sacks with the exact quantity directly after sorting. On the landing stage at Le Pavé, the customers are usually waiting to load their vans and trucks with baskets and sacks of mussels. They are there every day, Sunday included. Monday is a peak day for the delivery of mussels.

C. FARMING RISKS

C.1. Hydrographic conditions

Gales may spring up in the area, but serious damage to the bouchots is rare. When broken poles are found after a gale, one usually finds that shipworms are the primary cause of the damage. Neither low temperatures nor low salinities cause the mussel farmers any concern in this region.

C.2. Predators

A serious predator of mussels and other molluscan shellfish is the stingray *Trygon pastinaca*, locally called "tère". In order to limit the damage caused by this fish, which usually makes its appearance in great numbers from June onwards, the mussel farmers place wide-meshed nets, 100 to 150 m long, at the seaward side of the bouchot, parallel to the coastline.

The starfish *Asterias rubens*, a notorious predator in many mussel-cultivating areas, is not a serious nuisance here, as it does not thrive in intertidal conditions. Starfish encountered by the mussel farmers are never thrown back in the water, but are taken ashore and destroyed.

Amongst bird species, the diving scoter ducks (the "macreuses", *Melanitta nigra*) are feared by the local mussel farmers. These ducks are most numerous in September. In an effort to scare them away some mussel farmers attach long white strips of plastic, fluttering in the wind, on top of the poles of their bouchots. Hopefully these "languettes" will also keep away the brent geese *(Brenta bernicla)* and the seagulls, which feed on the small mussel seed. The oyster-catcher *(Haematopus ostralegus)* is said to do little damage here.

C.3. Other pests

The shipworm *(Teredo navalis)* is a notorious pest in the Pertuis Breton. The poles of pine or oak used in building a bouchot in due course become honeycombed by shipworms. They are thus much more easily broken. Usually the poles break just where they enter the sea bed. For this reason about 1/6 of all the poles have to be replaced every year. Impregnated wood would resist the shipworms for a number of years, but the results of experiments carried out by the local mussel farmers did not encourage them to use it on a large scale. *Teredo*-resistant tropical wood is not available in the desired dimensions and is very costly.

Other organisms causing inconvenience are the acorn barnacles (*Balanus* sp.) which settle on the shells of the mussels. Since the customers do not accept mussels with barnacles on their shells, the farmers have to eliminate them by rubbing the mussels thoroughly over the grid under a strong jet of water. A heavy settlement of barnacles means much extra work.

Other organisms the mussel farmers dislike are the hydroid *Tubularia larynx* and the sea anemone *Actinia equina*. Where the former abounds, mussel seed will not settle in appreciable numbers, whereas the latter may cause skin injury to the mussel farmers.

C.4. Parasites and diseases

In the mantle cavity of the mussel one may encounter the pea-crab *Pinnotheres pisum*, sometimes as often as one per 20 mussels. They consume some of the food collected by the mussels and should therefore be considered as a kind of parasite. The pea-crab is not considered to be one of the dangerous mussel enemies by the "boucholeurs", but the occasional ravages of this organism in American oyster districts should be taken as a warning.

The copepod *Mytilicola intestinalis* is considered a much more dangerous parasite. This organism was seen for the first time in the Aiguillon district in 1960. Most probably it had been introduced here accidentally a few years before in mussels originating from Brittany. Since *Mytilicola* has an adverse effect on the growth and fattening of the mussel, especially in the warmer season, it is considered as a serious pest. When the rate of infestation is high, byssus attachment becomes poor, and for mussels grown on bouchots, good attachment by byssus threads is of vital importance. Where great numbers of *Mytilicola* have invaded mussels, the latter have dropped from the poles and perished in the sediment. The studies of Brienne (1964) led to conclusions which tally very well with the experience gained in other mussel districts (Korringa, 1968): the rate of infestation is greatest close to the sea bed, since the *Mytilicola* larvae mostly seek the host there; the denser the mussel population, the greater is the number of parasites; the lower the current velocities and the lesser the degree of flushing, the poorer is the redistribution of the larvae, and the greater therefore is the percentage retained in the mussel area. Brienne's findings have led to drastic measures taken by the appropriate authorities to safeguard the mussel industry in the Aiguillon district: (1) a large reduction in the high number of illegal bouchots; (2) stricter adherence to the rules set for the distances between the bouchots; (3) total prohibition of "clayonnage" (interwoven bouchots); (4) a more rapid turnover rate established by the use of rope-collectors, by avoiding overcrowding, and by facilitating the flushing by the tide.

Though it was rightly considered impossible to eradicate *Mytilicola* completely by chemical or biological means, the measures mentioned above resulted in a lower infestation rate and prevented the loss of mussels caused by detachment from the poles on account of poor byssus attachment. Mussels lightly infested with *Mytilicola* may be of good quality for the market, but may however be less viable, and cannot therefore be transported over great distances.

D. ECONOMIC ASPECTS

D.1. Rents

M. Bouyé has to pay rent for 3 km of concession; for the larger part this is FFr. 0.20/m, and for his smaller concession, he pays the minimum rent of FFr. 50 per concession. The local "taxe professionelle", at 25% of the rents, is extra.

D.2. Inventory items

The most important inventory item is the 13 m long wooden motor boat equipped with a diesel engine of 75 h.p. The working life of the boat is estimated at 15 years, the engine, 7 years. Of the special equipment on board mention should be made of the radio-telephone installation, the washing and grading trough with the exchangeable trellis, the weighing machine, and the 3 h.p. petrol-engine pump. For proper maintenance the boat should be treated with marine paint four or five times a year, requiring 1 or 2 days work each time. The motor boat tows the 6 m wooden "yol" equipped with its 4 h.p. outboard engine. The yawl has an estimated life of 6 years. On board the motor boat are two lightweight wooden "acons" 3.10 m long, with a life of 5 years. For the storage of modest quantities of marketable mussels, the company uses two wooden racks which support wooden cages strung with metal wire netting. The wood has a life of 5 years, but the wire netting should be replaced every year. Ashore M. Bouyé uses a truck which has a carrying capacity of 500 kg and an estimated life of 4 years; this provides for local transport of poles and personnel, and is also used for transporting material to the Bay of St. Brieuc, on Brittany's North coast. The small tools such as scrape-nets, scrapers and shovels used on board, are less important items on the inventory list.

D.3. Expendable items

The poles of the bouchots last 5 or 6 years. Therefore successive purchases of oak poles from 4 to 6 m length, costing some FFr. 8 each, are an absolute necessity. M. Bouyé has 3 km of concession, used partly for the production of mussel seed (130 poles per 50 m) and partly for growing marketable mussels (84 poles per 50 m). The poles are bought from farmers or timber merchants in the area. An appreciable amount of capital is invested in the total number of poles the company has in the field. Therefore to enter the poles under the heading of expendable items instead of inventory items, is quite arbitrary. In procuring seed with coco-rope M. Bouyé uses 100 m of concession, which requires an annual purchase of 5 km of rope 1 cm in diameter, costing from FFr. 0.15 to 0.20/m, if bought directly from a wholesaler in Marseille.

In addition one has to purchase the cylindrical nylon netting to make the "boudins"; 10 m of netting are required for each pole of the growing area. The netting lasts about 1 year and costs FFr. 1/m. Since M. Bouyé reserves about 4 000 poles for growth and fattening, he has to buy about 40 km of netting each year.

For packing the mussels, M. Bouyé has to purchase sacks of woven plastic or of burlap with a capacity of 25 kg of mussels, and these are non-returnable. 4 000 sacks would be required if all the mussels were shipped in sacks. In fact, some are shipped in plastic baskets which are used repeatedly. Further, the company has to purchase fuel-oil for the 75 h.p. diesel engine of the boat. The boat is used almost every day; the journey to the nearest bouchot takes 1 h, and for those far away from the home port, 2 h. On arrival at the bouchots, one requires petrol for the outboard engine and the pump which is used for installing the poles and for washing the consumer mussels. Less important are the annual expenses for about 100 kg of iron nails, marine paint for the maintenance of the fleet, and rubber gloves.

D.4. Expected growth

Since the mussel seed is not bought but procured with the seed bouchots or with coco-fibre ropes, the amount of mussel seed one transfers to the growth and fattening bouchots is not recorded. If all the seed were put in the "boudins" it would be possible to estimate the amount, the "boudins" having a diameter of about 10 cm and a length of 2 × 5 m per pole, but for the seed collected with ropes it is more difficult to estimate the quantity. M. Bouyé expects to harvest each year an average of 25 kg of marketable mussels from every pole used for growth and fattening. This tallies with the fact that he uses about 4 000 poles for this purpose, and produces about 100 metric tons of marketable mussels per year.

D.5. Personnel

As has been stated above, M. Bouyé works full-time in the field, always assisted by one man. In times of high workloads, such as when building a new bouchot or making a rack for rope farming, he calls on the assistance of another mussel farmer. Mme Bouyé takes care of the administration, but is not on the pay-roll.

D.6. Sales

M. Bouyé produces about 100 metric tons of marketable mussels per year from those of his bouchots in the area under consideration. It should be noted here that he produces a further 50 tons on his 1.2 km of concession in the Baie de St. Brieuc on the North coast of Brittany. M. Bouyé goes frequently to St. Brieuc, especially between February 15th and April 15th, and hires workers there. The price the farmers receive for marketable mussels is fixed by the Syndicate in March each year. In 1971 the price for standard quality mussels, packed in sacks ("départ logé") was FFr. 1.35/kg.

D.7. Further development

M. Bouyé is not for the time being planning to increase the quantity of mussels produced. There is no shortage of mussel seed; if so desired he could easily step up seed production by using a greater length of coco-fibre rope. Availability of personnel is not a limiting factor either. The shortage of high quality concessions for rapid growth and fattening of mussels is the main restriction for the company under consideration. It is an economic disadvantage that M. Bouyé's concessions east of the river Lay are so very widely distributed. Re-allotment would certainly be welcomed by him. On the other hand the distribution of concessions over a wider area reduces to some extent the risk of damage by predators and parasites. The market for small-size consumer mussels does not seem to be flexible, least of all at the beginning of the season when the mussels will only tolerate transport over short distances.

E. GOVERNMENTAL ASSISTANCE

It will be clear from the foregoing that government agencies contribute much to the rational exploitation of the natural resources of the Pertuis Breton. Over-exploitation and overcrowding of the concessions are an ever-threatening danger where many are working close together, often in pursuit of a quick profit. The rules laid down for building bouchots, for maintaining the distances between bouchots, for the elimination of illegal installations, and for the prohibition of the traditional "clayonnage" system, are all worked out for the benefit of mussel farming in this area. The Institut Scientifique des Pêches Maritimes, which has a branch laboratory at nearby La Rochelle, keeps an eye on the mussel venture in this area and does not hesitate to come to the rescue with research results and advice when it deems necessary. The events which occurred after the introduction of the mussel parasite *Mytilicola intestinalis* speak for themselves.

F. GENERAL IMPRESSION

Mussel farming with bouchots, a semi-farming system with a very long tradition, has escaped the dangers of developmental stagnation. There has been no hesitation in introducing new methods and modern equipment, and thus developments have kept pace with the march of time. Still, this system of mussel farming is quite laborious and it cannot be expected that the mussel industry will in due course assume much larger proportions. The silting up of the Anse de l'Aiguillon, to which the mussel farming operations have certainly contributed for a very long time, made it necessary to build the bouchots gradually further and further away. The introduction of motorized vessels made this possible. Mussels do not grow to a large size here, presumably because they are farmed in the intertidal zone and because of physiological interference by excessive amounts of inert suspended matter. The minimum legal size is remarkably small (4 cm, formerly 5 cm), but the flesh is usually plump and of extraordinarily good flavour, and thus the regional markets eagerly absorb these mussels.

The total length of all the bouchots in the Aiguillon—La Rochelle area amounts to about 600 km, with an annual production of 7 000 tons of consumer mussels. The total mussel production in France amounts to about 45 000 tons, the main contribution to which (10 000 tons) comes from the hanging cultures in the Etang de Thau, comparable with mussel farming

Italian-style, whereas another 10 000 tons are fished directly from natural beds.

G. TECHNICAL DETAILS

G.1. The motor boat

The mother-vessel measures 13 m in length, has a beam of 3.50 m and a draught of 0.50 m. The hull is built of oak, and the deck, built 50 cm under the shearstrake, is made of Scandinavian pine. The wheel-house is placed aft, which is quite unusual for the mussel boats used in this area. The engine is a 75 h.p. diesel of the brand "Couach" (or English Ford). The boat is equipped with a radio-telephone to keep it in permanent contact with the headquarters at Charron. A pump, brand "Bernard", with a 3 h.p. petrol engine is used for installing the poles and for washing the marketable produce.

G.2. Tool for harvesting mussel seed from the poles

A sharp crescent-shaped steel blade 19 cm wide, 8 cm broad overall, is welded to a metal net 26 cm in diameter, 26 cm deep, consisting of 2 cm wide hexagonal meshes. The net is reinforced at its upper rim and at its deepest point by iron rods of 15 mm diameter. A long wooden shaft is attached to its ferrule (Fig. 47).

Fig. 47. Tools of the mussel farmer: harvesting scraper, with basket, shovel and scraper for cleaning the poles.

G.3. The "yol" auxiliary boat

All the mussel farmers in the area work at their bouchots with a manoeuvrable wooden "yol" (yawl). This is a simple sloop-type boat, a little under 6 m long and equipped with a 4 h.p. outboard engine. One can work from it standing upright, as in scraping the mussel seed from the poles for example.

G.4. Scrapers for cleaning the poles

The scraper consists of a simple crescent-shaped steel blade, 17 cm wide and 8 cm broad overall, with a ferrule 27 cm long to hold the long wooden shaft. It is used to scrape acorn barnacles and other unwanted growth from the poles of the bouchots once all the seed has been harvested and the poles have to be prepared for the following season (Fig. 47).

G.5. The nylon "boudins"

The nylon netting used to make the 5 m long "sausages" (boudins) is cylindrical, about 10 cm in diameter, with 4 cm meshes when stretched. Similar cylindrical netting is widely used for packing various foodstuffs. The netting need not be very strong, for the mussels will attach themselves in due course with their own byssus threads, and will crawl out of the meshes.

G.6. Acon

Very small light-weight boats are carried on the deck of the mother-vessel. They are still named "acons" after the mud-horses once used in this area. The acons measure 3.10 m in length. They are near rectangular in shape, 0.85 m broad fore, 0.90 m aft, with a depth of 0.45 m. They are made of 2 cm thick deal. In harvesting mussels from the bouchots by hand, one works in a kneeling position in the acon.

G.7. Shovel

The shovel which is used to transfer mussels from the acons to the motor boat has a concave blade 40 cm long and 33 cm broad at its lower end, curved towards the ferrule which holds the 1.20 m wooden shaft. The side edges of the blade are folded over.

G.8. Equipment for washing and grading on board

On the starboard side of the motor boat a wooden trough 2.20 × 0.45 × 0.28 m deep is installed. Its bottom consists of a metal trellis of 5 mm round iron rods with 11 mm spaces between them. For grading large-size mussels one can insert a trellis of 7 mm rods with interspaces of 13 mm. A broader trough can be installed for washing and grading mussels which are unusually dirty. At the front of the trough, part of the side wall is cut out to aid in filling the sacks with clean, graded mussels.

G.9. Storage cages ("arches" or "reservoirs")

Wooden frames supporting rectangular cages with bases and side walls of wire netting are used for the temporary storage of mussels. The cages are covered with wooden lids. M. Bouyé has two sets of such cages: one consisting of six cages of 2 × 1 × 0.50 m, the other of ten cages. The galvanized wire netting has a 16-mm mesh size.

G.10. Plastic boxes for packing marketable mussels

The boxes are rectangular and tapering, measuring 50 × 25 cm at the top and 32 × 25 cm at the base. Both the walls and the bottom are of strong plastic with large square perforations.

H. BIBLIOGRAPHY

Andouin, J., 1954. La mytiliculture en Baie de l'Aiguillon. Science et Pêche, Bull Inf. Doc. Inst. Pêches Marit., No. 16, 7—8.

Brienne, H., 1960. *Mytilicola intestinalis* Steuer dans les moules de la Baie de l'Aiguillon. Science et Pêche, Bull. Inf. Doc. Inst. Pêches Marit., No. 87, 6 pp.

Brienne, H., 1962. Evolution de l'infestation des moules de la Baie de l'Aiguillon par *Mytilicola intestinalis* Steuer au cours de l'année 1961. Science et Pêche, Bull. Inf. Doc. Inst. Pêches Marit., No. 106, 5 pp.

Brienne, H., 1964. Observations sur l'infestation des moules du Pertuis Breton par *Mytilicola intestinalis* Steuer. Rev. Trav. Inst. Pêches Marit., 28: 205—230.

Brienne, H. and Pairain, R., 1966. Distribution de *Mytilicola intestinalis* Steuer le long des côtes françaises. Science et Pêche, Bull. Inf. Doc. Inst. Pêches Marit., No. 149, pp. 1—6.

Callame, B., 1961. Contribution à l'étude du milieu meuble intercotidal (côtes charentaises). Trav. Cent. Rech. Etud. Océanogr., 4, I, II, III, 118 pp.

De la Blanchère, H. 1866. Culture des Plages Maritimes. J. Rotschild, Paris, 276 pp.

Fraiche, F., 1865. Guide Pratique de l'Ostréiculture et Procédés d'Elevage et de Multiplication des Races Marines Comestibles. Librairie Scientifique, Industrielle et Agricole, Eugène Lacroix, Paris, 176 pp.

Korringa, P., 1968. On the ecology and distribution of the parasitic copepod *Mytilicola intestinalis* Steuer. Bijdr. Dierkd., 38: 47—57.

Lambert, L., 1939. La Moule et la Mytiliculture. A. Guillot, Versailles, 55 pp.

Lambert, L., 1950. Les Coquillages Comestibles, Huîtres, Moules, Coquillages Variés. "Que sais-je?", 416. Presses Universitaires de France, Paris, 127 pp.

Lambert, L., Faideau, F. and Pluteau, R., 1969. Ostréiculture et Mytiliculture sur le Littoral Centre Ouest. Jean Foucher, La Rochelle, 78 pp.

Letaconnoux, R. and Audouin, J., 1956. Contribution à l'étude du pétoncle (*Chlamys varia* L.). Rev. Trav. Inst. Pêches Marit., 20: 133—155.
Lorin, J., 1970. Etude Sédimentologique de la Partie Orientale du Pertuis Breton et de la Baie de l'Aiguillon. Thèse 3e cycle, Bordeaux, 177 pp.
Marchand, H., 1914. Les procédés actuels de la mytiliculture en France. Ann. Soc. Linn., 61: 31—46.
Sarton, G., 1923. Walton, a mediaeval aquiculturist. Isis, 6: 307—310.

Chapter 4

MUSSEL FARMING IN GALICIA, SPAIN

Example
Industrias Maritimo Pesqueras, Proprietor-Director Don Luis Losada Lago,
Punta Preguntoiro (Villajuán), Villagarcía de Arosa (Ap. Post. 14).
Period of observation: 1970—1971.

A. BACKGROUND

A.1. General principles

Mussel farming Spanish-style is a semi-culture in which the first phase of the mussel's life-cycle, i.e. larval development, is left entirely to nature. The mussel seed settles freely on the rocks, especially in the intertidal zone, and during part of the year also on ropes especially hung for this purpose from the floating mussel parks. The mussel seed is collected by hand from the rocks and transferred to ropes which are hung from the specially-constructed mussel rafts. The seed which settles directly on ropes is in due course distributed on further ropes and then hung again from the rafts.

Hydrographic conditions prevailing in the rías of Galicia favour the production of a rich phytoplankton, allowing the mussels, which are continuously bathed in sea water, to grow rapidly. When the total weight of mussels on a given rope exceeds a certain limit, the ropes are taken up in order to distribute the mussels on to a greater length of rope. When the mussels have attained marketable size they are detached from the ropes and either sold to canneries or placed for some time in a purification basin to supply the markets at home and abroad with mussels of a high standard of cleanness.

A.2. Biology of the species

The mussel farmed in Galicia, Spain, is the common mussel, *Mytilus edulis* L., a general outline of the biology of which has been given in the chapter dealing with mussel farming in The Netherlands. Several authors claim that the mussels from the Mediterranean belong to another species, viz. *Mytilus galloprovincialis* Lamarck. There are certain features of shell shape and shell colour which seem to support that view, but on the other hand, *M. edulis* shows such a great variation in shell characteristics throughout its extensive geographical range that doubt can be expressed about whether the mussels from the Mediterranean are genetically different enough from those living in Atlantic waters that they should be assigned to another species. Some biologists claim that they can find specimens of *M. galloprovincialis* among the populations of *M. edulis*, as for instance in the coastal waters of Galicia, but for the time being, it seems wise to consider all the Galician mussels as a special population of *M. edulis.* Shell size, shell shape, and shell colour certainly deviate somewhat from those observed in the mussels farmed in The Netherlands and in France; the shells of the Spanish mussels are larger — up to 15 cm and higher, and they also reveal a more reddish tinge than the latter. The living tissues too show a distinctive aspect; they reveal a broader band of dark pigmentation in the mantle's edge, and in general the mantle itself is more brick-red in colour than that of the Dutch and French mussels. This latter feature, however, probably has a physiological rather than a genetic origin, for some carotenoids of the phytoplankton may accumulate in the mussel's mantle tissue. Light exerts a considerable effect on the exterior appearance of the shell, in particular on its degree of pigmentation. The more intense the illumination, the darker the shell tends to be. Andreu (1960) claims that a study of the pigmentation of the shell can reveal in detail the pattern of illumination under which the mussels have lived. According to Andreu, the light factor has appreciable influence on several aspects of the mussel's physiology. Growth, for instance, is faster in complete darkness, but under such conditions the mussels also produce fewer and weaker byssus threads. Andreu explains this in terms of greater wave action in higher water strata where the light is more intense, and a lesser tendency to be torn loose by waves in darker and deeper sites. Therefore mussels living under natural conditions simply have to attach themselves more firmly in the higher water layers. When the cultivated mussels are transferred to other ropes they may then be in water of a different light intensity, for the water of the Galician rías are so rich in phytoplankton that much of the sunlight is intercepted by it. According to Andreu, mussels adapted to a given light intensity will, when transferred, try fo find that same level of illumination and in doing so may detach their byssus threads. In mussel farming Spanish-style effective attachment of the byssus threads is of considerable importance, since mussels which, for whatever reason, detach themselves from the ropes can be considered as lost, as they will drop to the sea bed and perish in the very soft mud. Therefore Andreu advises that considerable caution be exercised in transplanting mussels, especially in the summer months.

A.3. Geographical situation

Mussel farming in Galicia in northwestern Spain is carried out in a series of inlets, called rías. Going from south to north these are the rías of Vigo, Pontevedra, Arosa (by far the most important in mussel production), Muros, La Coruña, Betanzos, Ferrol, Ortigueira and Vivero. The Ría de Arosa extends from 42° 27′ to 42° 40′ N and from 8° 56′ to 9° 01′ W (see Chart). The headquarters of the company under consideration, Punta Preguntoiro, is situated 42° 35′ N and 8° 48′ W.

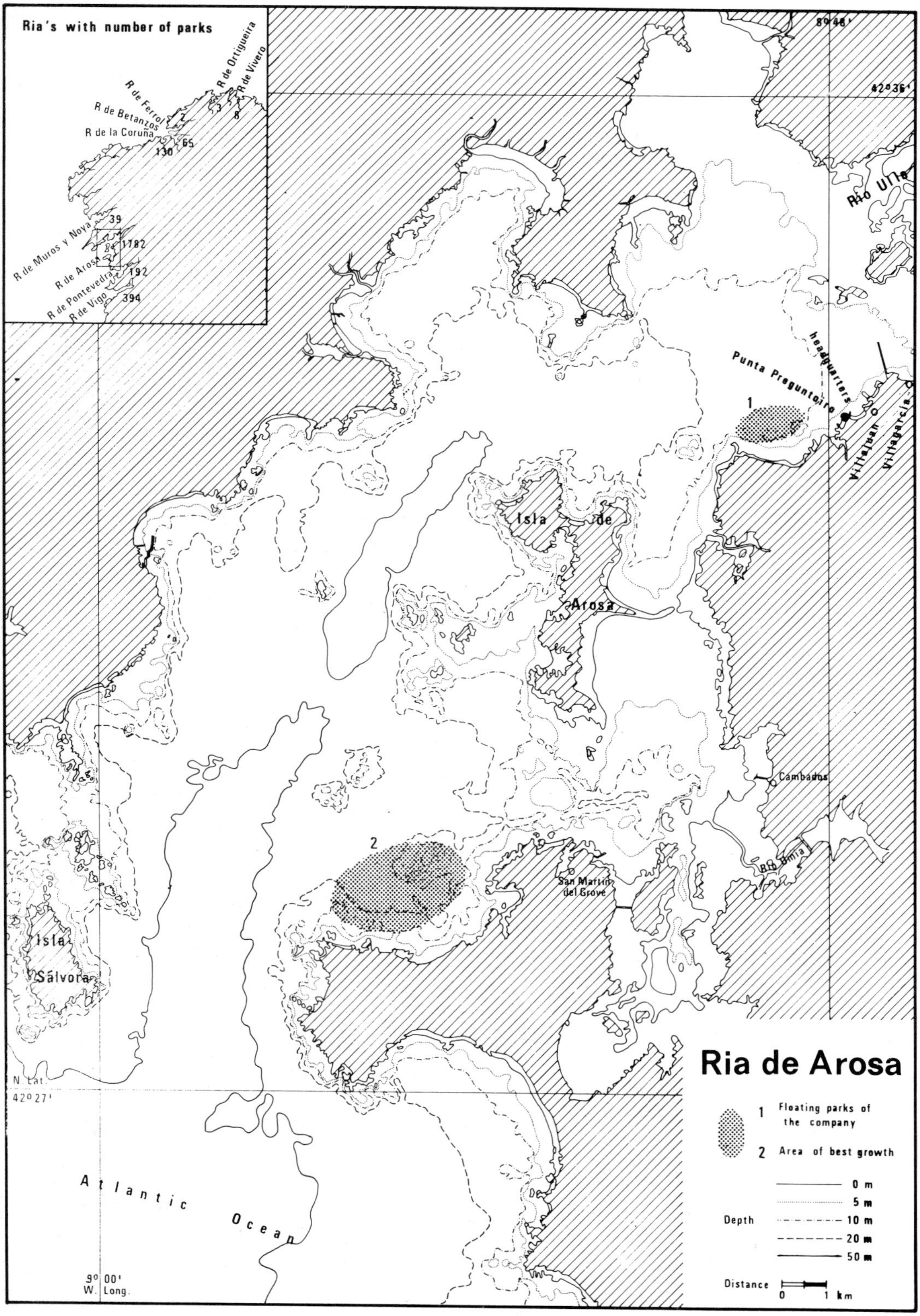

Ria's with number of parks
R de Ortigueira
R de Vivero
R de Ferrol
R de Betanzos
R de la Coruña
2
3
8
130
65
39
1782
192
394
R de Muros y Noya
R de Arosa
R de Pontevedra
R de Vigo
8° 48'
42° 36'
Rio Ulla
headquarters
Punta Preguntoiro
Villajuan
Villagarcia
1
Isla de Arosa
Cambados
Rio Umia
San Martin del Grove
2
Isla Sálvora
N. Lat.
42° 27'
Atlantic Ocean
9° 00'
W. Long.
Ria de Arosa
1 Floating parks of the company
2 Area of best growth
Depth
0 m
5 m
10 m
20 m
50 m
Distance
0
1 km

Of the four southern inlets ("rías bajas") in Galicia the Ría de Arosa is the largest. Its total length, from the Isle of Salvora at the entrance, to the estuary of the river Ulla at its upper end, amounts to 25 km in a straight line. The width varies between 3 and 12 km, on account of islands and coves; its total surface area is about 253 km^2.

A.4. Hydrographic pattern

Topography. The Ría de Arosa is, like the other "rías bajas", a deep bay, its central channel having a depth of about 60 m. By far the larger part of the bay has a depth of over 20 m. Silt, brought down by the river Ulla, has settled on the floor of the entire inner part of the ría as a soft mud. The outer part of the ría receives less sediment. Rocks exposed around the coastline of the bay consist for the greater part of granite, locally of gneiss and mica-schists.

Climate. In general the land surrounding the rías is high enough to offer a certain degree of shelter. The area has a temperate climate, with air temperatures of about 8° C in January and 20° C in July, and with about 1 500 mm of precipitation, falling mainly in winter.

Tides, currents and waves. In the inner part of the ría the tidal range amounts to 3.5 m at spring tides, and 1.1 m at neap tides, but strong westerly or easterly winds can affect these figures to such an extent that one has to add or subtract up to 0.40 m. The tidal currents show a maximum velocity of about 1 knot (50 cm/s) but the wind may exert a considerable influence, especially in the surface layers, so that the total current velocity, due to both tide and wind, ranges from 5 to 91 cm/s (Cadee, 1968). At 10 to 30 m depth the maximum current velocity is on average only 20 cm/s; close to the sea bed it is considerably less. The presence of soft silt on most of the bay's floor indicates that current velocities must be low there.

The waves of the Ocean may reach a height of 10 m on the Galician coast, but inside the bay the waves die away rapidly; 4 km from the entrance the highest waves are 7 m, at 7 km they are 5 m, at 13 km they are 2 m, and at 10 km from the entrance of the bay they are only 1 m (Cadee, 1968).

Water temperature. There is a marked difference in the range of water temperatures between the inner and the outer parts of the ría. In the outer part the winter minimum is about 13°C in a normal year and in the inner part, 11°C in the surface layer; the summer maxima are 17 and 21°C respectively. As usual under such topographical conditions the range is smaller near the entrance of the bay. The water mass in the Ría de Arosa is stratified during the greater part of the year, but reveals a period of homogeneity in the winter months. Therefore winter temperatures in the lower layers are almost the same — sometimes even slightly higher — than in the surface layers, whereas the summer maxima in the deeper layers amount only to 13°C both in the outer and in the inner parts of the bay. It is a peculiar phenomenon that the water temperatures in the deeper layers are often somewhat lower in summer than in winter! Northern winds prevail in summer, causing an upwelling of the shelf water outside the rías. This explains the low temperatures measured in the coastal water in summer. It is this cold water from outside which penetrates into the rías, and which accounts for most of the flushing.

Salinity. Since the influence of the ocean is the predominant one, the salinity fluctuates throughout the year between 35 and 36‰. Because the Ría de Arosa has, however, a rather large drainage area (3 560 km^2), an occasional spell of heavy rainfall may lead to a serious drop in the salinity of the surface layers, especially in the vicinity of the estuary of the river Ulla. Sometimes mussels bathed in these upper layers are killed by such freshets.

Oxygen. In the surface layers of the bay, in which the cultivated mussels live, one can expect the water to be saturated with oxygen; in the summer season a high oxygen production by the phytoplankton may even lead to slight supersaturation. In the deeper layers the water is usually undersaturated with oxygen since more oxygen is consumed than produced there. Particularly in summer, rather low values may be found in the deeper water layers, occasionally less than 50% of saturation level.

Plankton and turbidity. The water in the Ría de Arosa is very clear, as Secchi disc visibilities of 5 to 6 m indicate. Measurements of suspended matter yielded low values: 5 to 10 mg/1 according to Cadee (1968). The water is quite rich in phytoplankton, and the upwelling phenomenon described above ensures a regular influx of water rich in nutrients, which offers excellent conditions for diatoms and dinoflagellates. Measurements of the phytoplankton chlorophyll made in the inner bay near Villajuán by Figueras (1971) in the years 1966—1970 show that most of the time the amount ranges from 10 to 15 Harvey Units/l, but that peaks of about 40 Harvey Units occur regularly, and a maximum value of 90 has been recorded, e.g., early in the years 1968 and 1970. It is stated that one Harvey Unit corresponds to 1.30 μg N in the case of diatoms, and 2.50 μg N in the case of dinoflagellates (Vives and Fraga, 1960). Figueras and Andreu (1968) measured the quantity of suspended organic matter in the years 1966 and 1967 and showed that this fluctuated around 8 μg-at N/l during most of the year, with occasional peaks up to 15 μg-at N/l. The primary production shows an annual average of 0.9 g C/m^2/day, ranging from 0.07 to 7.6 g C/m^2/day (Andreu, 1968, referring to a communication of Fraga). For photosynthesis, the following figures are given: annual average 10.5 μg C/l/h, ranging from 1.1 to 61.6 μg C/l/h. In mid-summer, especially in the month of July, the phytoplankton is less rich. It is most probably the continuous influx of cold water welling up and mixing readily with the upper water layers of the bay during winter, which largely governs the fluctuations in the phytoplankton pattern.

Sea bed sediments. As has been stated in the section "Topography", the inner part of the bay has predominantly a soft mud bed. In these deposits over 90% of the particles are under 50 μm in diameter (Cadee, 1968). Coarser sediments are found not only in the outer part of the bay and on the beaches, but also in the shallow section southeast of the Isla de Arosa. Since mussel cultivation as carried out in the Ría de Arosa takes place in the superficial water layers only, the nature of the bed deposits is not of crucial importance.

A.5. Legal aspects

Official permission is needed to moor a mussel raft ("vivero flotante") in the Spanish rías, and this permission should be requested from the Ministerio de Comercio in Madrid. In doing so, a design must be submitted of the raft that is to be built, indicating its dimensions, its construction, and the material to be used, with a concise explanation of its intended use (e.g., for mussel farming or for oyster farming). The Ministry seeks advice from various directions to avoid a clash with other interests such as navigation, public works, recreation, and national defence. But the health department (Sanidad Nacional) must be approached to ensure that the site chosen does not suffer unduly from sewage pollution, and also the fisheries officials of the Instituto Español de Oceanografía should be consulted, to ensure that the planned increase in the production of mussels keeps pace with the gradual expansion of the market, and to avoid over-exploitation of the section of the ría under consideration.

Once permission has been obtained, one is not required to pay a lease to the government. The instructions given should be observed, as, for instance, the minimum distance of 100 m from other rafts. Where navigation is of some importance (e.g. in the Ría de Vigo) the rafts have to be moored with two anchors to keep them in position throughout the changing of the tide.

B. SEQUENCE OF OPERATIONS

B.1. Procuring the seed

Mussel seed settles abundantly on the rocks of the intertidal zone in the coastal waters of Galicia. Due to the great population densities in such sites and the high level at which the larvae prefer to settle, such mussels will never grow and fatten well enough to yield a marketable product. This natural resource was not therefore exploited for human consumption until mussel farming began.

Various sources are tapped to procure the mussel seed required for farming. In general, the young mussels which settle naturally on the rocks still constitute the most important source of all, yielding 60 to 70% of the seed used for farming in Galicia. The company under consideration obtains about half its supply of seed from this natural source, the remainder being collected with the aid of ropes hung from the floating parks.

In collecting mussel seed from the rocks, a boat is used to transport both the workers and the seed. If mussel seed occurs in abundance nearby one uses a simple rowing boat, manned by two people, but if a good supply of seed has to be obtained from further off, one of the larger motor boats [1] which are used in mussel farming is taken; a rowing boat is also taken. Usually the seed is most abundant at the seaward ends of the rías. Particularly rich sources of mussel seed are the Cies Islands and adjacent rocks at the entrance of the Ría de Vigo. Others are the Island of Ons in the mouth of the Ría de Pontevedra and the Island of Salvora. Mussel farmers working in the Bay of Arosa often go to these places to collect mussel seed. To harvest the seed, one man jumps onto the rocks and scrapes off the seed with a "rasqueta", a spade-like steel blade usually about 15 cm square, with a wooden shaft of about 1 m length. The detached mussel seed is collected in baskets or other containers, which are emptied into the boat. This work can only be carried out when the water is calm. Spring or summer is the best time for the harvesting of seed from the rocks. When the water is too high, or where the mussel seed has settled somewhat deeper, another tool, the "rastrillo", an iron rake of about 30 cm width with teeth 10 cm long and 2 cm apart and a long wooden shaft is used. To this rake a piece of netting made of synthetic fibre is attached to catch the mussel seed scraped loose from the rocks.

Though the farmers are dependent on tides and weather conditions, and have to sail a considerable distance to find the seed (the Cies islands are some 35 km away from Villajuán), they often manage to collect up to 1 500 kg of seed per tide, working for about 4 h on site with rasqueta and rastrillo. But only good sailors, well-acquainted with the vagaries of the sea, will regularly collect such an amount.

The seed harvested either can be used as primary material for one's own floating parks or it can be sold to other mussel farmers. The price fetched is about P 3.50/kg. Generally it is economically advantageous to buy seed for that price, for hired labour means an expense of P 300 to 350 per person per day, and the weather does not always allow the collection of 1 500 kg of seed. The company under consideration usually buys the seed it needs rather than sending out its own workers to collect seed.

A second way to procure mussel seed is to collect it on ropes hung from the mussel rafts [3]. One can use ropes made from different materials. In the bay of Vigo ropes plaited from esparto grass *(Stipa tenacissima)* are normally used, these being attached horizontally under the rafts in the upper water layers. The company under consideration employs the more usual mussel ropes made of synthetic fibre, hung vertically from the rafts with a weight (stone or iron) of about 3 kg at the lower ends to keep them stretched. Great numbers of such ropes are hung out in December and January to catch the seed in February and March. Ropes are hung out until autumn, but the best catches of the seed are invariably made in spring.

It has been frequently observed that mussel seed may settle profusely on the rocks in autumn, but it is not clear why settling on the ropes is nearly always poor in autumn. It can hardly be a result of a scarcity of larvae when good settling occurs at that time on the rocks. Perhaps the lapse of time between hanging out the ropes and the attachment of seed, parallel to the experience gained with the "bouchot" culture in France, can give the explanation. Mussel seed prefers to settle primarily on filamentous objects such as hydroids. It may be surmised that the ropes gather a more adequate primary coating of filamentous organisms in early spring than later in the year, but only careful scientific investigations can give the answer to this question.

A third source of seed can also be found on the rafts. The ropes, which bear mussels of various sizes, usually receive a thick coating of seed just under the water level. This "cuff" of seed at the upper end of the rope is easily harvested by hand by kneeling on the beams of the raft and lifting the ropes. For the company under consideration this source of seed is of some importance. The seed that occasionally settles in between the large mussels is, however, not used here as a primary source for farming, since it requires too much manual labour to separate it.

B.2. Attaching the seed to the ropes

Once the mussel seed has been procured from the rocks it has to be attached to the ropes without delay. Whenever possible it should be hung from the rafts the same day.

Various types of material are used for making mussel ropes. Originally rope plaited from crude or retted esparto grass (*S. tenacissima*), produced in eastern Spain, was used for this purpose. This is quite a cheap material (P 18/kg), but it is not very strong and it has a rather short usable life. It can be strengthened with a nylon cord, although this increases the cost, or its life-span can be extended to 3 years by a thorough impregnation with tar, a job which is dirty and unpleasant. Therefore other materials have superceded the esparto grass. Excellent ropes are plaited from four narrow strips cut off old trawl nets (4 cm meshes) made of synthetic fibre. They are strong and rough, offering the mussels an excellent hold. Such ropes are quite cheap, have a life-span of up to 7 years, and are therefore widely used, but the material is not always readily available. Another cheap material that can be used as a mussel support consists of square or flat section strips of rubber, cut from old tyres, but the company under consideration uses almost exclusively ropes made of synthetic fibre, which are strong and have a very long life. There are two types in use, a smooth one of blue colour, and a more woolly type resembling manila, also excellent for catching seed. This material costs P 50/kg (1 kg is about 8 m), lasts a long time, and is always available.

The length of the individual ropes varies from 6 to 9 m, according to the depth of the site in which the park is moored, for the ropes should never touch the muddy bed. Usually the ropes are 7 to 8 m long. Some experiments are made with ropes of double length, but to avoid too heavy a load, only the lower part of such ropes is used for mussel growth. A loop is spliced in both ends to facilitate attachment and detachment of the rope to the floating parks, and to simplify turning the rope when the mussels grow faster in one section than the other, thus ensuring a product of uniform size.

Important features of mussel ropes are the "palos", pegs of eucalyptus wood (measuring 25 × 1.5 × 1.5 cm and costing P 0.20 each; usable life, 4 to 5 years) which are inserted in the ropes by twisting them open at intervals of about 40 cm. If in due course the ropes carry a heavy load of mussels (e.g., 50 kg) the latter may slip off in stormy weather or during the hoisting of the ropes for harvesting. The "palos" prevent such disaster by taking the weight of the mussels in the section of the rope immediately above.

Fig. 48. Walking slowly backwards one attaches the mussel seed to the ropes with "techido de malla", a bandage-like tissue specially made for the Spanish mussel industry. (Photo P. Korringa.)

To attach mussels to the rope, the rope is stretched horizontally between two stands at a height of about 1.15 m. The stands are separated by a distance corresponding to the length of the rope, e.g., 9 m. A metal hook on top of each stand serves to fasten the rope. At one side a set of pulleys may be used to pull the rope taut. The stands are constructed either on the mussel rafts themselves, a convenient place for the transfer of mussels from one rope to others (thinning out), or ashore in a shed giving protection against the weather. The company used here as an example, employing female hired hands for this phase of the operations, has chosen the latter solution.

The mussels to be attached to the ropes, either as seed or at the half-grown stage, are placed on a wooden trolley of simple construction, which can be dragged along with one foot while the operator walks slowly backwards along the rope, attaching the mussels to it. With one hand one takes some mussel seed and with the other one holds a roll of tissue which is wound around the seed. This bandage-like tissue is called "techido de malla" (tissue with meshes) and is specially made for the Spanish mussel industry [2]. It takes about 6 or 7 min for an experienced worker to load a 8 m rope with seed (Fig. 48). As estimated by eye, from 1.5 to 1.75 kg of seed is attached per metre of rope.

If the seed is of small size (under 1 cm shell-length) one uses less per metre of rope, and if one is attaching half-grown mussels one rope takes more, e.g., 3 kg/m.

B.3. Hanging the ropes loaded with seed from the rafts

When the seed has been fastened to the ropes these should be brought to the rafts without delay, whenever possible the same day. They are transported to the floating parks [3] (Figs 49 and 50) on the "barca", the 12 m motor boat in use by the Galician mussel farmers [1]. The loop at the end of the rope serves to fasten it to a special piece of thinner polyester rope (12/14 mm thick), the upper end slung around one of the girders of the raft. This piece of rope suffers more wear and tear, being partly in water and partly in air, and can be renewed whenever necessary. In practice it will not last longer than 3 to 4 years. Another advantage of the "loop" system is that the ropes carrying the mussels can easily be turned over when so desired, the aim being to produce mussels of equal size. The largest parks of the catamaran type carry about 1 000 ropes, but the usual type of mussel raft, measuring about 18 × 18 m, carries from 600 to 700 ropes. The ropes are attached in such a way that the mussels will be exposed to an adequate stream of water carrying their food, and also so that there will be no risk that the ropes will touch each other in rough weather. This is achieved by hanging 2 or 3 ropes per m^2 of raft. It is important to note here that it is regular practice to use the floating parks as efficiently as possible. This means that each park usually carries a variety of ropes; ropes for collecting seed, ropes coated with seed for pro-

Fig. 49. Floating park for mussel farming provided with a shelter in the Ría de Arosa. On such a park the artisan farmer does all the work himself. (Photo P. Korringa.)

ducing half-grown mussels, and ropes for growing marketable mussels. Thus the rafts are continuously productive. If one operates with a single raft only, one is, of course, compelled to use a diversity of ropes; the more rafts in use by a given company, the more one can specialize if so desired.

B.4. Growth

Since the mussels are continuously bathed in water which is rich in phytoplankton and small detritus particles throughout the year, and since an adequate degree of flushing is ensured by the tides, feeding conditions are optimal. The mussels grow rapidly in the Galician rías, particularly as winter temperatures are never so low that growth comes to a complete standstill; at most growth slows down somewhat in mid-winter. The "techido de malla" serves only to keep the mussels together during the first few days in irregular positions around the central rope. Soon they crawl through the meshes of the gauze, detaching and re-attaching the byssus threads, in order to gain the optimal position for feeding, siphons facing outward. The remains of the bandage can later be found underneath the mussels. Interesting observations have been made by Fraga and Vives (1960) on the retention of plankton and

Fig. 50. Large floating park with the hull of an old barge as a centre piece. This type of park is eminently suited on poorly-sheltered sites. (Photo P. Korringa.)

detritus by the mussels during passage of the sea water through the forest of ropes hanging from a mussel raft. They analyzed water samples taken at 2.5 m depth in front of, and behind, the raft, and concluded that from 35 to 40% of both phytoplankton and detritus were taken up by the mussels during a single passage of the sea water underneath the mussel raft. Apparently mussels do not discriminate between living phytoplankton and detritus in their filter-feeding. Though the water contains almost three times as much detritus as living phytoplankton the assumption that detritus is the more important item in the diet of the mussels will only be supported when it can be demonstrated that both materials are equally well digested and have the same food value.

Considering the remarkably high degree of retention of particulate organic matter by the mussels of one single park and taking into account the rapidly increasing number of rafts moored in the Ría de Arosa (according to the latest figures the number of floating parks here has exceeded 2 000) it is not surprising that the growth of the mussels is now slower than in the past when the number of rafts was much smaller. In this earlier period it was quite normal for the mussel seed to grow to the marketable size of 8 to 10 cm shell-length in 8 to 9 months. Close to the entrance of the bay growth is fastest; in the area west of El Grove, for instance (see Chart) it is still possible to produce marketable mussels within 8 to 9 months, but the rafts are considerably less sheltered there than in the inner part of the bay, and for the company under consideration this site would be too distant from its headquarters. In the central part of the Ría de Arosa nowadays it takes 12 to 14 months between the seed stage and harvest. The type of seed used is also a factor to be taken into account; seed collected from the rocks usually shows a more rapid growth than that obtained from ropes hung from the rafts, and the seed which settles in autumn requires fewer months to mature than that which is established in spring, since the latter suffers from the decline in growth during mid-winter.

B.5. Thinning out

When hung out in spring the mussel seed may grow so fast that it becomes advisable to thin it out 3 or 4 months later. Seed hung out in autumn can be left longer, up to 6 months. A coating of mussels which is too dense would not only lead to a slower growth, but would also increase the risk that clusters of mussels will fall off in rough weather. Therefore the ropes should be taken up when the load reaches about 10 kg of mussels/m, and proceed with the thinning out ("desdoblar"). In the smaller private enterprises the ropes may be hoisted up by hand and the transfer of the half-grown mussels to other ropes may be done on the raft itself (Figs 51 and 52), but this company takes up the ropes mechanically [4] and transports some 50 ropes at a time on the motor boat to headquarters where the ropes are unloaded [5] and brought into the large shed. The half-grown mussels are scraped from the rope by hand with the aid of a simple iron hook ("gancho"), and scooped into boxes. Attaching the mussels once again to ropes proceeds along the same lines as has been described for the attachment of seed to the ropes, using ropes stretched between two stands and a trolley for carrying the mussels. Again this work is done by female workers sheltered from the weather. About 3.5 kg (estimated by eye) of half-grown mussels are attached per metre of rope, so that the mussels detached from one rope are now re-distributed on 2 or 3 ropes. Whenever possible, they are brought back to the raft on the same day. It becomes possible during the thinning-out operations to eliminate by hand any fouling organisms which compete for space and food, such as ascidians, which may have settled on and in between the mussels. Now follows a period of about 8 to 10 months, sometimes up to 12 months, during which the mussels require little or no handling at all. One simply has to wait till the mussels attain the marketable size of 8 to 10 cm shell-length. It may, however, happen that one notices a dangerously heavy load of mussels on the ropes. It is then advisable to go to the trouble of taking the ropes up again to re-distribute the mussels onto a greater length of rope. If one

Fig. 51. Formerly the floats were equipped with tall masts to hoist up the ropes carrying mussels when they needed to be thinned out or harvested. Photo taken in the Ría de Vigo in 1957. (Photo P. Korringa.)

wishes to take the trouble of producing mussels of closely similar size one should turn the ropes over during the growing period.

B.6. Harvesting the marketable mussels

Marketable mussels should not only be of a given shell-length, but should also have good quality edible parts. If a large percentage of any batch of mussels is on the verge of spawning or has just spawned, the farmer will wait till the mussels are in better condition for the market. Here, however, it is

Fig. 52. It often happens that small mussel seed settles in between the full-grown mussels. It takes a considerable amount of manual labour to sort it out. The seed can be used to coat new ropes. (Photo P. Korringa, 1957.)

impossible to restrict the sale of mussels to a part of the year that is completely free from spawning activity. Because of the narrow range of water temperatures, a certain percentage of the mussels contains either eggs or sperm at any time of the year. The limited degree of strictly seasonal spawning activity is the reason that Spain does not have an official close season for the sale of mussels, though there is a noteworthy seasonal fluctuation in the sales, with a slump early in summer, coinciding with a peak in spawning activity in May. The mussels are considered of excellent quality when about 50% of their total weight can be ascribed to the raw soft tissues, 25% to the shells, the remaining 25% being sea water. If the mussels are boiled first, the

Fig. 53. Modern mussel boats are equipped with a power crane and metal basket to collect the ropes with half-grown or fully-grown mussels. (Photo P. Korringa.)

figures are different: 25 to 30% of the weight can then be ascribed to the cooked edible tissues. Top qualities can be expected from August to December.

In harvesting, the same technique is used as was described for taking up ropes with half-grown mussels [4]. Again, two ropes can be hoisted simultaneously and caught in the metal basket straight under the bow of the boat (Fig. 53). As many ropes as can be handled ashore are taken up.

B.7. Cleaning and grading

On arrival at the shed, the mussels can be removed from the ropes simply by shaking them off, or by hand-picking; when they are strongly attached by byssus threads the gancho is used. One can expect to harvest about 10 kg of mussels per m of rope. Depending on the length of the ropes (6 to 9 m), the number of ropes per raft (600 to 1 000) and the growing time required from seed to marketable mussels (10 to 16 months), one raft may produce from 30 000 to 90 000 kg of mussels per year. The mussels are still clustered together, and it is up to the farmer to decide what to do next. Often the mussels are graded and cleaned in the boat by rubbing them over a grid of iron bars and washing them thoroughly. Thus one separates the mussels and eliminates the smaller ones together with silt, empty shells and other unwanted objects. These mussels, packed in nylon bags, are ready to be sold on the home markets or to the mussel canneries. This company follows a different system, since the major part of its product has to be prepared for export. Breaking up the clusters in the way described involves tearing loose the byssus threads. This is such a physiological jar for the mussels that they will often not survive such treatment. Sometimes, a machine equipped with saw-blades to cut the byssus attachments is used, but the result is then worse if one has to keep the mussels in storage for some time prior to shipment. Declustering should be done very carefully, preferably by cutting the byssus threads with a pair of scissors, but this operation would be economically prohibitive.

The company under consideration keeps those mussels destined for export in clusters, but eliminates by hand all mussel seed, empty shells, ascidians and other unwanted objects. This is time-consuming work, carried out by female workers, handling about 200 kg of mussels per person per 8-h working day. All the mussels the company buys from other farmers should undergo the same treatment, since declustering on board with the metal grid renders them unsuitable for storage.

B.8. Storage and purification

The company of Don Luis Losada has constructed huge storage basins [6], some of which are indoors. It is probably the largest mussel storage plant in the world, with 800 m^2 of indoor basin plus 1 200 m^2 of outdoor storage basin (Figs 54 and 55). The water is pumped a distance of 30 m through iron pipes of 35 cm diameter, from a depth of 3 m under the surface at low water level. Before the water is poured into the large 4-m deep settling tanks, a dosage of sodium hypochlorite (1—3 ppm) is injected to ensure a high standard of bacteriological purity. The water near the headquarters is not always sufficiently pure, since some domestic sewage enters the bay from Villajuán. In the winter season, extra care must be taken because land drainage may then

Fig. 54. The plant "Industrias Maritimo Pesqueras", of Don Luis Losada Lago, which is probably the largest installation for cleaning, storing and shipping mussels in the world. (Photo courtesy Don Luis Losada Lago.)

carry some muck and farmyard manure spread by farmers on their fields. No industrial pollution need be feared here. A pharmacist-bacteriologist gives 6 h of his services per day in this purification plant to ensure the highest standard of purity of the mussels shipped from here.

The 70 cm deep indoor storage basin (Fig. 56) has a sloping floor and a series of longitudinal walls 20 cm high at the upper end and 35 cm high at the lower end, on which baskets containing mussels can be placed. The pumps can pass 700 m^3/h of sea water from the settling basin through this storage basin. The baskets each contain 17 kg of mussels. The company uses 6 000 plastic baskets plus 2 000 iron ones. The baskets with the mussels are placed by hand on the walls and taken out the same way. For bacteriological safety it suffices to keep the mussels 24 h in the storage basin, but to be absolutely sure, they are kept there for at least 48 h. Mussels destined for export are kept in the basin for about 2 weeks and are periodically exposed to the air. Such treatment reduces the loss in quality during transportation and reduces the risk of spawning after packing. In the warmer season, one should refrain from exposing the mussels in the open basins to air during the daytime.

Fig. 55. Don Luis Losada Lago in front of the settling basin of his mussel storage plant. In the background the pumps and the pipelines leading to the basin. (Photo P. Korringa.)

The storage basins have a maximum capacity of 80 000 kg of mussels. If space is available, the company offers other farmers the possibility of purifying and storing their mussels, at the price of P 3 000 per 1 000 kg.

B.9. Shipment and marketing

For shipment the mussels are packed in sacks of gauze made of a synthetic material, each holding 15 kg (Fig. 57). Yellow bags are prescribed for the shipment of purified mussels. The bags are closed with a metal wire turned tightly with a special spiral-shafted hook. In the warmer season temperature-

Fig. 56. The indoor storage basin is filled with purified sea water when the baskets with mussels have been placed on the longitudinal walls. (Photo P. Korringa.)

controlled trucks are used to transport the mussels to the border, e.g., to Irun. Chip baskets containing 6 or 12 kg are often used for shipping mussels to Paris, Bordeaux and Toulouse.

About 80% of the mussels packed by this company are exported, about 20% go to the home markets, and no mussels from this company are sold to the mussel canneries. By far the larger part of the mussels shipped by the company is purchased as marketable mussels from other farmers, viz. about 95%, produced by some 200 rafts; this means that only 5% come from Losada's own floating parks. 90% of the purchased mussels come from parks in the Ría de Arosa, the remainder from installations in the Ría de Ponteve-

Fig. 57. Without employment of a great number of female hands, the present Spanish mussel industry could not survive. A drastic increase in wages would seriously endanger its remunerativeness. (Photo P. Korringa.)

dra and the Ría de Vigo. The total exported product of the company amounts to about 3 000 metric tons per year.

C. FARMING RISKS

C.1. Hydrographic conditions

Though the salinity of the Ría de Arosa is usually high, with a very limited range of fluctuation, excessive rainfall may lead to some freshening of the surface layers once in 6 to 7 years, especially in the inner part of the bay where the river Ulla discharges. Under such conditions it may happen that the mussels on the topmost sections of the ropes (0.5 to 1 m) die off. Sublethal effects such as an interruption of growth can also result from a drop in salinity. Such events occurred for instance in March and April of 1963.

Every winter gales cause some damage to parks on exposed sites. When leaks occur in the floats, the entire park may sink. If they are not brought up very soon afterwards, the mussels on the ropes may perish in the soft mud of the bed, but in any case one should expect a loss of 30% of the mussels when a raft sinks. In the winter of 1970—1971 an exceptionally strong gale, lasting several days, sank 25 mussel rafts in the Ría de Arosa.

C.2. Biological conditions

Considering the fertility of the rías, one wonders whether excessive blooms of certain dinoflagellates would not once in a while render the mussels dangerous to the consumer. Strangely enough such a phenomenon is of very rare occurrence here and it is many years since the Spanish biologists have observed a bloom of noxious phytoplankton in the Galician rías.

C.3. Predators

Since the mussels are kept off the sea bed they suffer very little from the usual enemies such as crabs and starfish. Occasionally the starfish *Marthasterias glacialis*, the boring gastropod *Nucella lapillus* and the sea-urchin *Paracentrotus lividus*, which gnaws the edges of the mussel shells, are seen on the ropes. They have either settled there as planktonic larvae or they may have crawled up the anchor chain. However, the damage done by these predators is as a rule, very limited.

More dangerous is the sparid fish *Diplodus sargus* which may occasionally cause havoc among the mussels, crushing their shells. Some farmers protect their rafts against the ravages caused by these fish by fencing them in with netting.

C.4. Parasites and diseases

Though the mussel parasite *Mytilicola intestinalis* does occur in the Galician waters, it is only found in very limited numbers in the intestinal tracts of the mussels living on the ropes. At the head of the Ría de Vigo somewhat higher numbers have been observed, but never so high that the growth and quality of the mussels are seriously affected. Since the larvae of *Mytilicola* have a natural tendency to seek their host on the sea bed, a hanging culture escapes the destruction this creature brings about elsewhere.

No mussel diseases have been recorded in Galicia since mussel farming was started there in 1946.

C.5. Competitors

Fouling organisms such as the ascidians *Ascidiella aspersa* and *Ciona intestinalis* are competitors for space and food. They are eliminated by hand when the mussels are thinned out. No chemical treatment is applied. In a sense, mussel seed settling in between the larger mussels should also be considered as a fouling organism, competing for space and food. They too must be eliminated by hand in the cleaning procedure, which takes time. Incrusting organisms such as balanids and the polychaete worm *Pomatoceros* sometimes settle on the shells of the mussels in appreciable numbers. Their presence on the shells of marketable mussels is undesirable, but little can be done to keep them away.

Considerable caution is required with regard to shipworms *(Teredo)*. They do not affect the mussels as such but can destroy wooden structures in a remarkably short time. Therefore the mussel rafts should be protected against the attacks of shipworms. Experimental mussel farming in Venezuela has shown what can happen when this factor is overlooked or when the right material is not locally available.

C.6. Pollution

The events at La Coruña in November 1970 may sound a warning for other mussels districts. The coaster Erkowit, loaded with pesticides among which were dieldrin and mercury compounds, got into trouble on its way to the Sudan and was finally shipwrecked in the bay of La Coruña, releasing part of the poisonous cargo. All the mussels growing on the floating parks in the area had to be destroyed and it was more than a year before farming could be resumed.

C.7. Theft

Since theft of mussels from the rafts has never occurred here, special supervision is not deemed necessary.

D. THE ECONOMICS OF MUSSEL FARMING IN SPAIN

D.1. Rents

As has been noted above, the Spanish mussel farmers need not pay any rent for mooring their floating parks in the rías.

D.2. Inventory items

The company under consideration operates six floating parks for mussel farming, one being a sturdy old wooden boat with a system of outriggers built around it, very convenient for an exposed site, and the other five being large rafts of the catamaran-type, carrying about 1 000 ropes each. It costs P 500 000 to build such a catamaran-park, whereas the more usual type of park with four floats and carrying 600 to 700 ropes, costs P 300 000. A well-constructed raft has a very long life, counted in tens of years.

Another important inventory item is a motor boat. Such a boat, equipped with a 24 h.p. diesel engine, costs about P 250 000 and has a life of about 20 years. Since no heavy work such as dredging has to be done, the engine too has a long life. The powered crane on the bow costs about P 50 000.

As will be clear from the foregoing, the purification and storage basins ashore are out of proportion to the quantity of mussels farmed by the company. The company markets about 20 times the quantity of mussels it produces with its own parks. The size and construction of the basins, shed, pumps and pipelines may give some idea of the magnitude of the investment. It will, however, be interesting to mention the cost of some apparatus that would also be necessary for a smaller plant: the mussels are dragged from the quay into the shed on a wagon with pneumatic tyres (P 15 000) by a 10 h.p. electric motor (P 20 000) and a 50 m steel cable 12 mm thick. The plastic baskets in use in the storage basins cost P 160 each, the iron baskets P 100.

D.3. Materials

The purchase of ropes is a recurring item on the budget of the Spanish mussel farmer. Rope made of synthetic fibres such as that used by this company costs about P 50/kg (8 m) and though it has a life of many years, it should be replaced occasionally. Each catamaran raft carries 1 000 ropes of 6 to 9 m length. A further item is the "techido de malla" which costs P 25 per roll of 200 m, sufficient to load 6 to 8 ropes with mussels. It is used once only, its service being required during the period that the mussels are left to grow on the ropes (3 to 6 months from seed to half-grown mussels, 8 to 12 months from half-grown to marketable size). Sometimes a third quantity of tissue is required when extra thinning out is deemed necessary.

This company buys about half the seed it needs. About 100 boxes, each holding 40 kg of seed, at a price of about P 3.50/kg, are required to load the ropes of one catamaran raft adequately.

Other recurring items are fuel and lubricating oil for the motor boat, maintenance of boats and engines, electricity for lighting, electric motors and pumps, and sacks for packing the mussels (P 4.50 for a gauze sack which can hold 15 kg of mussels).

D.4. Labour and personnel

Since it is not easy to separate farming work from the work required in storage and shipment, and since the latter involves a much larger quantity of mussels than the former, it should only be mentioned that this company employs 15 male labourers full-time throughout the year, and that up to 40 female workers are employed in the winter, and about 20 in summer. The wages are in the order of P 250 to 350 per day. The men do the work at sea, the women remain ashore in the Arosa district. It can be deduced from the description of the techniques employed how much time is required for loading the ropes with mussels, transferring mussels to a greater length of rope, sailing to and from the floating parks, and harvesting marketable mussels. Cleaning and grading is without doubt the most time-consuming operation.

Special mention should be made of the pharmacist-bacteriologist who is hired for 6 h per day to supervise the standard of cleanness.

D.5. Purchase and sale of marketable mussels

The company buys large quantities of marketable mussels from other farmers, in total the production of about 200 parks. These mussels still have to be cleaned, graded, purified and stored. The price the company has to pay for these mussels (up to P 5/kg), depends not only on size and quality, but also on the current market price in Spain. The company sells about 3 000 tons of mussels per year, 80% for export, and 20% for the home markets. The price of export mussels is P 8/kg, pre-packed in sacks and carriage paid to the border. One should reckon on P 1/kg for transport to Irun in an ordinary truck, and P 1.30 for temperature-controlled transported. Transport to Irun takes about 20 h; the mussels are reloaded on to French trucks in Hendaye.

D.6. Further development

From the fact that the company under consideration buys far more marketable mussels than it produces from its own floating parks, it could be tentatively deduced that it is more profitable to buy marketable mussels from small private farmers — who do not count their hours — than to produce mussels with hired labour. However, the fact that Mr Losada has submitted an application for the construction of six more parks, contradicts such an assumption.

It cannot be denied that mussel farming Spanish-style requires a considerable amount of manual labour and that mechanization has not been widely introduced as yet. Therefore, the cost of labour determines to a very great extent the production costs for marketable mussels. The price of export mussels has not changed appreciably in recent years and since the price fetched for mussels produced in France, The Netherlands and Italy has to be taken into account, no more than P 7/kg including freight to the border, can be asked.

The fertility of the rías is so high that it is surely possible to increase the number of floating parks still further. The price the consumer is willing to pay will, together with the local labour situation, determine how much further mussel farming will develop in Galicia.

E. GOVERNMENTAL ASSISTANCE

The Vigo Laboratory of the Instituto Español de Oceanografía devotes much of its attention to mussel farming in the southern rías, whereas the branch laboratory at La Coruña does likewise for the mussel district further north. In addition, the biologists of the Instituto de Investigaciones Pesqueras of the Consejo Superior de Investigaciones Scientificas, which also has a laboratory at Vigo, carry out many investigations on the biology of the mussel and on some aspects of mussel farming. The conclusion of the biologists is that the Ría de Arosa is so rich in plankton that it would in principle be possible to moor up to 1 000 more floating parks there without running the risk that growth would decrease alarmingly, or that diseases and parasites would take advantage of the denser population. Additional parks should preferably be moored in the deeper, less-sheltered parts of the ría, where the mussels would be able to collect their food from the deeper water layers. It is important that the Spanish government guides the Galician mussel industry in such a way that the increment in the production keeps pace with the expansion of the markets at home and abroad.

F. GENERAL IMPRESSION

Mussel farming in Galicia is a striking example of rational exploitation of a natural resource which could not be put to use without man's intervention. Early in the century, a system of mussel farming was developed on racks on the Mediterranean coast, particularly in the areas of Tarragona and Barcelona, based on the example of the old Italian system of mussel farming. It was the biologist M. Oliver of the laboratory of Mallorca who suggested to the industrialist Don Alfonso Ozores Saavedra, Señor de Rubianes and Marqués de Aranda, that he might initiate an experiment with floating mussel parks in the Ría de Arosa. This initiative led to the construction of the first park in 1946, closely followed by one of Fernando Molins in the Ría de Vigo. It was this spirit of enterprise combined with scientific advice which led to a rapid development of the Galician mussel industry. Fortunately it was possible to

Fig. 58. The hundreds of mussel rafts moored in the Ría de Vigo give the impression of an armada ready for invasion. (Photo P. Korringa.)

develop gradually a market for mussels in parts of Spain where mussels were thus far an unknown item of the diet.

Initially old barges were used for the construction of mussel parks, but by the early nineteen fifties special mussel rafts were constructed. Step by step the materials and methods were improved. Ropes of esparto grass have now almost completely been replaced by ropes of synthetic material. The specially-developed crane at the bow of the mussel boats recently rendered the long masts on the rafts superfluous and eliminated much heavy labour. The work which was originally carried out on board the rafts, predominantly by female labourers and in all kinds of weather, now largely takes place, at least in the Arosa area, in the shelter of the sheds ashore.

Since mussels destined for export should keep their valves tightly closed during transport, they are repeatedly exposed to air for 1 to 2 weeks. Formerly, this was done simply by putting them on the beach, a procedure which involves serious risks in warm weather, but now Galicia possesses about 12 storage basins for this purpose.

Now the total mussel production from the 3 000 or more parks in Galicia lies somewhere between 140 and 160 million kg of marketable mussels. Of this roughly 10% are exported, especially to France and Italy, about 50% are brought in the fresh state to the home markets, and about 35 to 40% go to the mussel canneries, which make, among other things, the famous "escabeche". Part of the canned mussels finds its way abroad, for example to Germany.

The possibility of further mechanization, the price of labour in Galicia, and the further development of the markets at home and abroad, will be the main factors governing the future of the Galician mussel industry.

G. TECHNICAL DETAILS

G.1. The "barca"

The "barca", the motor boat (Fig. 59) used for mussel farming in Galicia, weighs 9 tons, is about 12 m long, 3 m broad, has a draught of 60 cm and 50 cm freeboard. It is constructed locally of 3 cm thick native pine ("pino gallego"). It has a deck in front and a small wheel-house aft. This company's boat is equipped with a 24 h.p. diesel engine of the brand "Samofa". On the bow is the crane for lifting the ropes loaded with mussels, and this is powered by an auxiliary diesel engine of 12 h.p. of the brand "Dieter". Some other mussel boats in the area use the same engine for both propulsion and for their crane.

G.2. "Techido de malla"

The "techido de malla" (Fig. 60) used by the company is of the brand "La Vieira", gauge E 20, available in rolls of 200 m length, width 18 cm, with nine meshes across. It is a loosely-woven tissue of which the warp is thicker and stronger than the weft. It is made of synthetic fibres of 20 to 24

Fig. 59. The boat of the modern Spanish mussel farmer is equipped with a power crane to lift the ropes with mussels. A sorting table with grid is used to wash and grade the marketable mussels.

Fig. 60. The "techido de malla" is specially made for the Spanish mussel industry and serves to attach the mussels to the ropes. (Photo P. Korringa.)

μm thick, two intertwined strings of 38 threads each for the warp, and one bundle of 38 threads for the weft. The meshes can easily be stretched so that the mussels can crawl through the tissue. It costs P 25 per roll and one roll of 200 m suffices to load six to eight ropes of 8 m length with seed. For very small seed one uses another gauge of tissue 14 cm wide with 12 meshes across.

G.3. Floating parks

The mussel rafts of the catamaran-type used by the company (Fig. 61) rest on two floats of 16 m length, triangular in section, each float consisting of three compartments. The floats are constructed of 4 cm thick local pine (pino gallego), but sometimes of pine imported from Scandinavia. In making the floats one follows quite closely the construction techniques of boat building. To protect the wood against shipworms (Teredo) and, in the higher parts of the floats, against the gribble (the isopod *Limnoria lignorum*), one nails 1.5 × 2 cm laths all over the float allowing 20 cm interspaces. On these laths is fixed metal gauze with a mesh size of about 3 cm. Next, a thick coating of concrete (1.5 to 2 cm) is applied through and on the metal gauze, and

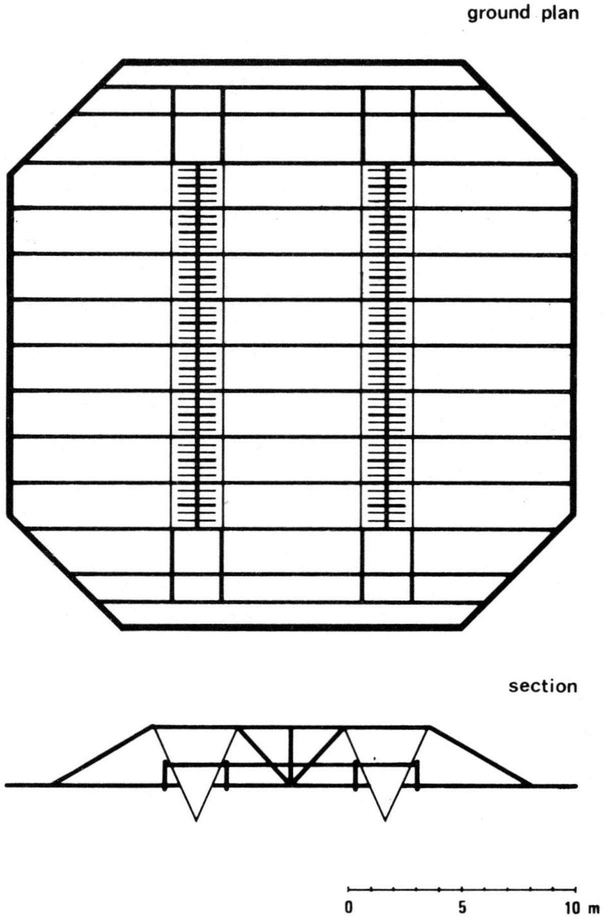

Fig. 61. Construction of a floating mussel park of the catamaran type.

this concrete protects against the shipworms. For newly-constructed rafts a thin layer of fibreglass will be used instead of concrete on wire gauze. Since it sometimes happens that a leak springs in the float in very rough weather, with some risk of the raft sinking, the floats of newly-constructed rafts are filled with expanded polyester to avoid this risk.

On the floats a superstructure is built, the "emparillado" of which the principle beams ("maestros") are made of eucalyptus of 30 × 30 cm and 22 × 20 cm. Next, 6 × 8 cm beams of the same wood are placed in between the main beams. The latter are interspaced at 60 cm intervals, and are the beams to which the mussel ropes will be attached, 60 cm apart. The catamaran rafts are strengthened by a system of wooden girders over the top of the floats.

In other types of raft there are either iron floats (Fig. 62), cylindrical in

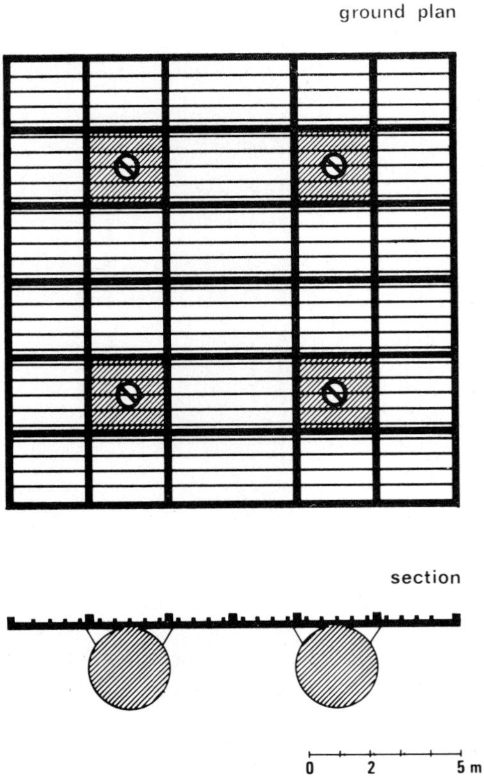

Fig. 62. Construction of a mussel raft with four iron floats.

shape (2.10 m diameter, walls 8 mm thick and well tarred), or cubical wooden floats (e.g., four floats of 2.5 to 3 m side or, in other cases, one centrally-placed float of 12 × 4 × 2 m) protected against shipworms in the same way as described above for the catamaran.

In the Ría de Arosa the floating parks are moored with one single anchor chain made of 35 mm thick iron, of a length six or seven times that of the local depth, attached to a very heavy weight ("muerto"). The floats turn round with the tides. Therefore a swivel ("giratoria") is inserted about 1 m under the surface of the water. Mussel farmers wishing to carry out all work at sea, construct the stands for the ropes and a shelter for protection against rainy weather on their rafts.

G.4. Mechanical lifting of the mussel ropes

In the older parks, long masts were installed for lifting the ropes, heavily laden with mussels, by hand. The motor boats in the Ría de Arosa are all

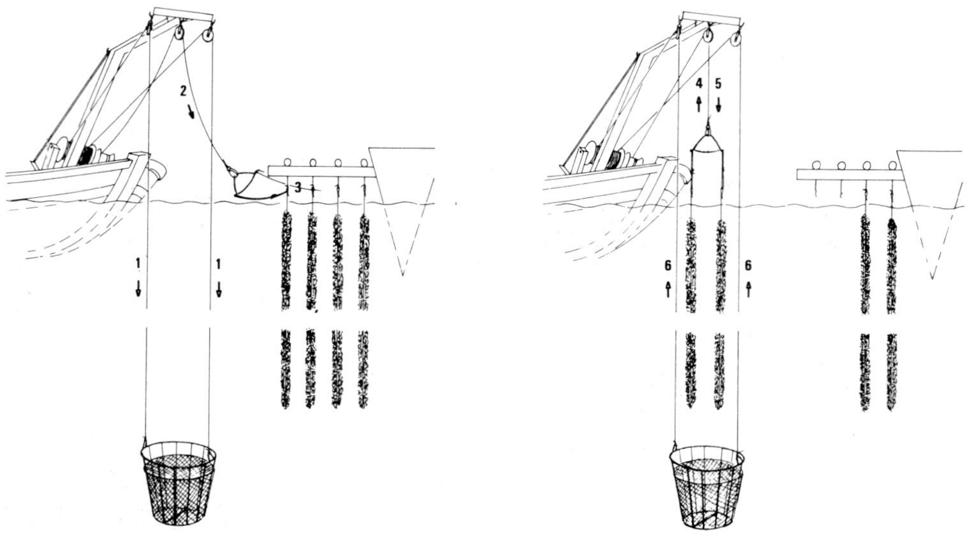

Fig. 63. Mechanical lifting of mussel ropes with the aid of a power crane.

nowadays equipped with a power crane, installed in the bow. The cranes have three pulleys, the two lateral ones are used to control a large metal basket which is lowered onto the bottom of the sea when the boat has taken up position at the site of the park from which mussels are to be harvested. The central pulley is provided with a long thin steel wire which can be taken out to the park. It is passed by hand underneath the beams of the park and then two mussel ropes are attached to it by the end loops. Then the ropes are detached from their supporting beams, allowing them to sink away sideways. At that moment the central wire is pulled in somewhat till the upper parts of the mussel ropes emerge. The ropes then hang vertically from the crane and when the wire is slackened off, they drop into the metal basket directly under the bow of the boat. By winching the lateral wires the basket emerges, containing the two coiled-up mussel ropes. The basket is then swung in till it is above the front deck. One movement with a handle opens the bottom flaps so that the mussel ropes drop onto the deck. After detaching the rope and swinging the basket outboard again to lower it to the sea bed, one is ready to take up the next couple of ropes. (See Fig. 63.)

G.5. Mechanical transport at the headquarters

On arrival at the quay, the mussel ropes are loaded onto low wagons running on four pneumatic tyres, and with a load-bearing surface of 200 × 90 cm. These can be dragged with a cable, wound by a 10 h.p. electric motor, into the shed where the mussels will be cleaned and graded.

G.6. *Purification and storage basins*

Some interesting details of the installations for storage and purification are as follows. Water can be pumped in by two 50 h.p. electric pumps with a capacity of 700 m^3/h, working independently day and night. The 30 m iron pipelines are constructed of sections of 5 m length, connected with flanges so that they can be taken apart for cleaning when mussel seed has settled too densely inside the pipes. As a rule, this cleaning should be done at least once a year. The hypochlorite is kept in 100 l plastic jars and exact metering is performed with a special membrane pump made by Hildebrand, Bad Ems.

A good flow of water, preventing stagnation in corners and near the bottom, is ensured by the gentle slope of the bottom of the 800 m^2 indoor basin. In this basin the baskets of mussels, measuring 60 × 40 × 20 cm, are placed some distance above the bottom on longitudinal walls, and extra flushing is also achieved by forcing the water to take a deeper course underneath a transverse wall, to be led over the top of the next wall in a series of jets. The water is not re-circulated, but let out into the bay again after having passed through the storage basin.

H. BIBLIOGRAPHY

Andreu, B., 1958. Sobre el cultivo del mejillón en Galicia. Ind. Pesq. (Vigo): 745—746, 44—47.

Andreu, B., 1960. Ensayos sobre el efecto de la luz en al ritmo de crecimiento del mejillón *(Mytilus edulis)* en la Ría de Vigo. Bol. R. Soc. Esp. Hist. Nat., Secc. Biol., 58(2): 217—235.

Andreu, B., 1962. Ordenación de la explotación Marisquera A. Aspectos cientifíco y técnico del cultivo del mejillón y industrialización de la ostra en Galicia. Ponencia Pesca, presentada al il Consejo Económico Sindical Provincial, Vigo (Pontevedra), 1962, 3—10.

Andreu, B., 1963. Propagación del copépodo parásito *Mytilicola intestinalis* en el mejillón cultivado en las rías gallegas (NW de España). Invest. Pesq., 24: 3—20.

Andreu, B., 1965. Biologia y parasitologia del mejillón gallego. Las Ciencas, Madrid, 30: 17—18.

Andreu, B., 1968. Pesquera y cultivo de mejillónes y ostras en España. Publicaciónes técnicas de la Junta de Estudios de Pesca, Núm. 7, Madrid, pp. 303—320.

Andreu, B., 1968. Fishery and culture of mussels and oysters in Spain. Proc. Symp. Mollusca, Mar. Biol. Assoc. India, Part III, pp. 835—846.

Cadee, G.C., 1968. Molluscan biocoenoses and thanatocoenoses in the Ría de Arosa, Galicia, Spain, Zool. Verh. Rijksmus. Nat. Hist. Leiden, 95: 121 pp.

Canel Garcia, 1968. Mejilloneras. Rev. Ingeniería Naval, Madrid, pp. 1—7.

Figueras, A., 1971. Parque experimental de ostricultura de Villajuán. Invest. Pesq., 47: 181—210.

Figueras, A. and Andreu, B., 1968. Estudio de los factores ambientales, crecimiento y mortalidad de la ostra plana y del ostión. II Periodo 1966—1967. Publicaciónes técnicas de la Junta de Estudios de Pesca. Núm. 7, Madrid, pp. 275—302.

Fraga, F., 1956. Variación estacional de la composición quimica del mejillón *(Mytilus edulis)*. Invest. Pesq., 4: 109—125.

Fraga, F., 1959. Relación entre peso, talle y composición quimica en el mejillón *(Mytilus edulis)*. Invest. Pesq., 14: 121—127.

Fraga, F., 1960. Variación estacional de la materia organica suspendida y disuelta en la Ría de Vigo. Influencia de la luz y de la temperatura. Invest. Pesq., 17: 127—140.

Fraga, F., 1967. Hydrografía de la Ría de Vigo, 192, con special referencia a los compuestos de nitrógeno. Ind. Pesq., 31: 145—160.

Fraga, F. and Vives, F., 1960. Retención de partículas orgánicas por el mejillón en los viveros flotantes. Reunion Prod. Mar Explot. Pesq., Barcelona, 4: 71—73.

Gallego, J.G., 1971. Estudio de la condiciónes oceanográficas en la Ría de Arosa, en verano. Biol. Inst. Español Oceanogr., No. 147, 39 pp.

Lubet, P., 1957. Cycle sexuel de *Mytilus edulis* L. et de *Mytilus galloprovincialis* Lmk dans le Bassin d'Arcachon. Ann. Biol., 33: 19—29.

Otto, L., 1975. Oceanography of the Ría de Arosa (N.W. Spain). Meded. Verh. Kon. Ned. Meteorol. Inst., Publicatie 102, No. 96, 210 pp.

Sanchez y Sanchez, M., 1956. Cultivo del mejillón en las Rías Gallegas. Ind. Pesq., 30: 697—698.

Chapter 5

MUSSEL FARMING IN THE ROADS OF LA SPEZIA, ITALY

Example
Cooperativa "La Proletaria", Viale S. Bartolomeo 525, La Spezia.
Period of observation: 1971—1972.

A. BACKGROUND

A.1. General principles

Mussel farming Italian-style is a semi-culture in which both larval development and the provision of food for juvenile and adult mussels is left entirely to nature. Though the mussel farmers sometimes use mussel seed which settles naturally on rocks and other solid substrata as primary material, by far the larger part of the seed they need is procured with the aid of artificial collectors.

Characteristic of the Mediterranean, are the very narrow tidal range and low current velocities. In the sheltered area where mussel farming is practised the sea bed is predominantly soft and muddy. Therefore the mussels must be grown above the sea bed. So Italian mussel farmers construct mussel parks called "vivai" — networks of poles connected with horizontal ropes. From these ropes they hang their strings of mussels which are permanently bathed in sea water and do not touch the muddy layer. A special type of rope is stretched around the parks to collect the seed. To control fouling organisms, the strings of mussels are exposed for some time to air whenever necessary.

Most of the work is done at sea. The marketable mussels are cleaned and graded and either sold in strings or neatly separated and packed, ready for the kitchen. The centres of consumption are the Italian urban areas.

A.2. Biology of the species

The edible mussel of the Mediterranean is usually named *Mytilus galloprovincialis* Lamarck, but as has been discussed in the chapter on mussel farming in Galicia, it differs from the Atlantic species *M. edulis* so little that there is no practical reason to present a separate treatise on the biology of the Mediterranean mussel. Its biology and external appearance are virtually the same as those of the Atlantic species, and it is a purely academic question whether one should really separate the two species. For the purpose of farming and sale, the Italian mussels can be considered as fully comparable with those farmed in the Atlantic waters.

A.3. Geographical situation

Along the Italian coast one finds several areas where mussels are farmed in hanging culture. In order of importance can be mentioned the area of Chioggia—Venezia—Trieste, the region of Taranto, the Gulf of La Spezia, the area of Napoli (including the Lago's Fusaro, Miseno and Lucrino) and also the regions of Olbia and Varano. The total Italian production of marketable mussels fluctuates between 25 000 and 30 000 metric tons per year.

The example chosen is the area of La Spezia, facing the Ligurian Sea (see Chart). The city of La Spezia, where the cooperative "La Proletaria" has its headquarters, is 44° 7′ N and 9° 48′ E. The city has been built at the head of an inlet of rectangular shape, 4.5 km long, 3.5 km wide, with a depth of 10 to 12 m. This inlet is called the Roads of La Spezia (Rada di La Spezia) and is ideally sheltered to the west by a mountainous peninsula, and further by the Isle of Palmaria. It is not suprising that this inlet is used as a harbour for both the merchant fleet and for the Italian navy. To improve the situation still further, a breakwater, the Diga Foranea has been constructed running approximately SW—NE across the entrance of the inlet, allowing passage at both ends. Most of the mussel parks are constructed in the well-sheltered area to the northwest of the breakwater, some others in sheltered sites north of Porto Venere and on the northern shore of the Isle of Palmaria.

Part of the mussel seed which the farmer uses as primary product is collected outside the Roads of La Spezia on sites where it settles naturally on the rocks; such places are usually located somewhere in the Gulf of La Spezia, e.g., at Tellaro, at Punta Bianca and at the south side of the Diga Foranea.

A.4. Hydrographic pattern

Depth. Though the Roads of La Spezia generally have a depth of 10 to 12 m, the sites where the mussel racks are built are as a rule close to the shore, where the depth is only 5 to 8 m, with the exception of the important parks along the Diga Foranea, where the depth is 11 to 12 m.

Tides. The tidal range is small throughout the Mediterranean and the effect of the wind on the water level is often of greater importance. In the Roads of La Spezia the maximal difference between high and low water caused by wind and tides amounts to about 30 cm. Current velocities are rather low where tides do not exert their influence. The waters of the Gulf of La Spezia are, however, not stagnant. Micro-currents of a rather irregular pattern cause a certain degree of flushing. Current velocities are in the order of 0.1 to 0.2 knots.

Waves. The geographical situation offers ideal shelter in the Roads of La Spezia, and because of the Diga Foranea waves are of limited height even in stormy weather.

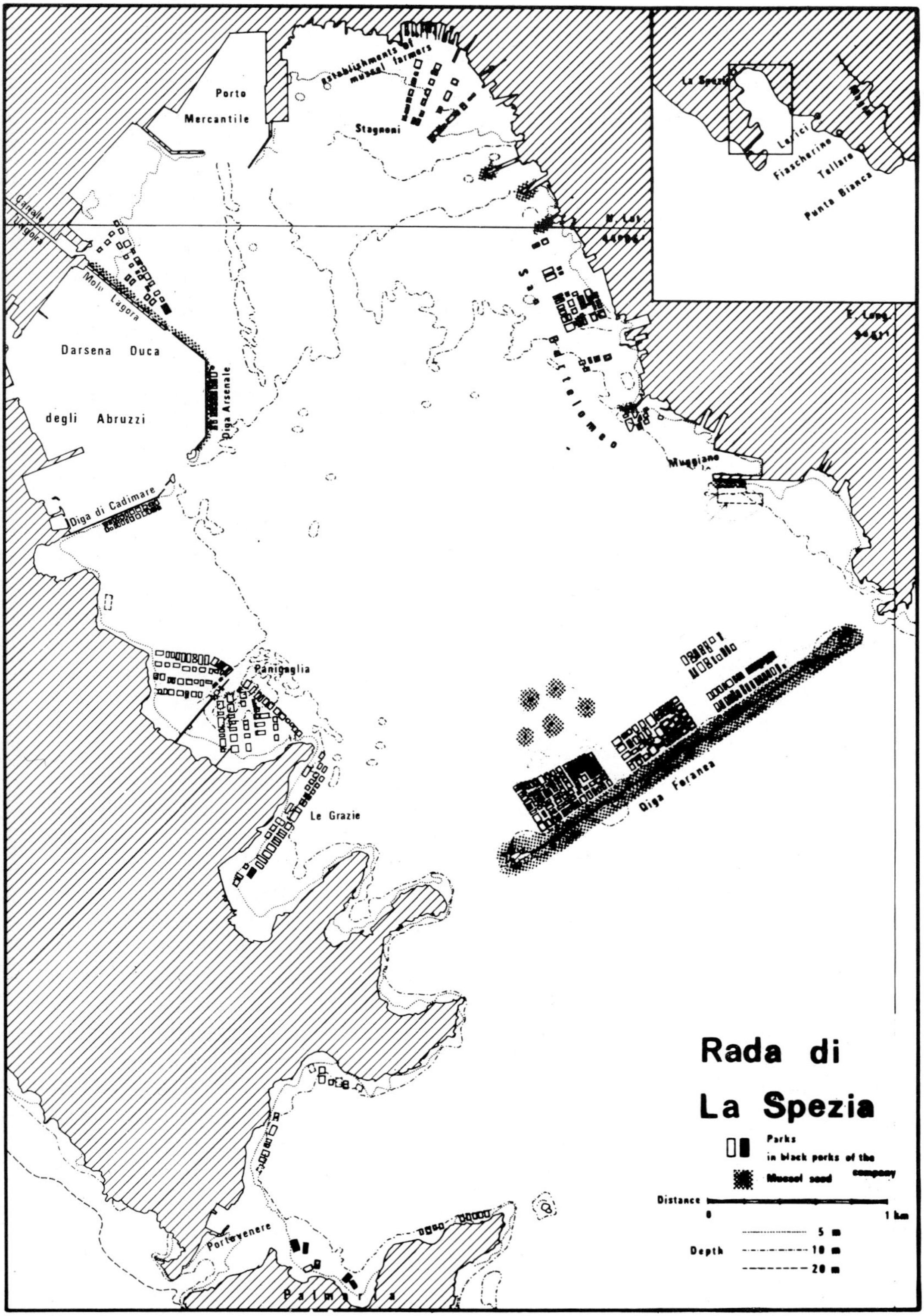
Rada di
La Spezia
Parks
in black parks of the company
Mussel seed
Distance
0
1 km
Depth
5 m
10 m
20 m
Porto
Mercantile
establishments of mussel farmers
Stagnoni
Canale Lagora
Moli Lagora
Darsena Duca
degli Abruzzi
Diga Arsenale
Diga di Cadimare
San Bartolomeo
Muggiano
Panigaglia
Le Grazie
Diga Foranea
Portovenere
Palmaria
La Spezia
Lerici
Fiascherino
Tellaro
Punta Bianca

Water temperature. In winter the water temperature may drop to 11 to 12°C in January, to rise to about 16°C in March and April, reaching a summer maximum from 23 to 25°C in July and August. The water is quite homogeneous; stratification does occur in the warmer months but is only observed offshore with a thermocline at a depth of 20 to 30 m. In October the thermocline is found as deep as 60 m; it disappears soon after.

Salinity. The salinity is very stable in the area and fluctuates between 37 and 38‰. The river Magra discharges south of the Gulf of La Spezia, but its influence is hardly noticeable in the Roads of La Spezia.

Plankton. Both phytoplankton and zooplankton have been studied extensively in the Gulf of La Spezia by staff members of the Laboratorio per lo Studio della Contaminazione Radioattiva del Mare at Fiascherino, situated about 10 km southeast of La Spezia. Quantitative data on the biomass of the phytoplankton (diatoms, peridineans and Coccolithophorids) reveal marked fluctuations with peak values occurring in the winter months and sometimes late in summer. Expressed in mg dry weight/m^3 the yields are: March—May 1.2 to 20.7, June—August 1 to 11.8, September—November 0.4 to 25.3 and December—February 3.4 to 29.2 (Nassogne, 1972).

Observations in the Roads give figures of the same order of magnitude, on average about 20 mg fresh weight/m^3, corresponding to 8×10^{-8} mg chlorophyll A/m^3 according to Zattera (personal communication) using a conversion factor of 4×10^{-9} to relate phytoplankton fresh weight to chlorophyll A. One should realize, however, that these figures have a relative value since they refer to recognizable phytoplankton organisms only and that both small flagellates and detritus particles which may be of paramount importance as food for mussels are not included. Empirically, it is clear that there is enough food in the water at La Spezia to ensure normal growth of the mussels.

Pollution. Since the Roads of La Spezia is a harbour of considerable importance and since several sewers of the city discharge into this body of water, it is not surprising that pollution is a serious problem. Nevertheless oysters were cultivated till they were wiped out in 1919—1920 by an epidemic of a microscopic parasite, and mussels were farmed thereafter. Though mussels grown in the area of influence of sewers cannot meet standards of bacteriological safety, the latter can be achieved by placing the mussels prior to marketing in a clean park near the Diga Foranea. Though pollution has always been a matter of concern in this area, the local mussel farmers claim that the situation has deteriorated since about the year 1960. On the north-northeastern shore an electric power plant of 2 000 MW (ENEL) has been constructed to replace the older one of 600 MW. This results in thermo-pollution of a serious nature and some groups of mussel parks had to be abandoned in that area. Mussels may still be killed by warm water in the summer months in the nothern part of the Roads.

Apart from incidental cases of pollution of a chemical nature (1962) and by oil (1968) there are industries such as the methane works in the north-west of the area and industrial waste discharged through the Lagora Canal in the same section which arouse concern.

The mussel farmers of La Spezia claim that they did not only lose formerly productive mussel parks, but that on the whole the mussels grow slower than in the period prior to 1960, and have deteriorated noticeably in quality and flavour in recent years, for which the farmers blame water pollution. To minimize the risk, they keep their mussels further above the sea bed than in former times and concentrate their activities more on the parks away from the danger spots. But exploitation of deeper parks, situated further away from headquarters and requiring longer poles, reduces the profitability of their enterprise.

Sea bed conditions. The sea bed of the Roads of La Spezia consists of thick layers of a soft black mud and the water immediately above it is of poor quality for mussel farming. Oil residues and other chemical wastes may be accumulated in the deposits. Since mussel farming makes use of the mid-water layers only, the nature of the substrata is not of crucial importance.

A.5. Legal aspects

The company under consideration is a cooperative called "La Proletaria". This cooperative consists of 18 participant-proprietors, who share equally in the net profit, one of whom, Mr Mario Corsini, stays ashore as head of the administration, responsible for the accountancy and the marketing of the products. Though all the other participants work at sea, the cooperative finds itself compelled to hire some extra hands on a full-time and seasonal basis.

For construction of a mussel park, official permission is required, which will only be given for some well-defined sectors in the area. It is the "Ministerio della Marina Mercantile" which leases sections of the Gulf to mussel farmers, who have to contact the "Capitaneria di Porto del Compartimento Marittimo di La Spezia". In total there are over 360 plots in the area leased to 80 enterprises, amounting to a total area of 234 000 m^2. Since there should be wide corridors between the mussel parks, for which no lease need be paid, the total area reserved for mussel farming amounts to 780 000 m^2. The cooperative "La Proletaria" has at its disposal 31 plots with a total area of 27 000 m^2, excluding corridors, distributed in nine groups. These plots vary in size from 500 to 2000 m^2; six of their plots are situated outside the Roads on the northern shore of the Isle of Palmaria, and measure from 1 000 to 2 000 m^2. In addition, the cooperative has at its disposal 1 025 m^2 for landing stages, storage space and sheds at the headquarters ashore.

The official lease to be paid is L. 40/m^2 of water and L. 80/m^2 of land. Since the Italian Government wants to stimulate exploitation by cooperatives, all the company has to pay per year is a rent-charge of L. 5000, irrespective of the area it really uses.

There is no regular police patrol on the parks. Therefore the mussel farmers of La Spezia have organized a surveillance system for themselves for which the cooperative has to pay L. 58 000 per month. There exist special printed forms to report theft of mussels to the police at the port of La Spezia, but owing to the presence of the paid watchmen it is rarely necessary to use these forms.

B. SEQUENCE OF OPERATIONS

B.1. Procuring the seed

The mussel farmer can collect seed, sometimes even half-grown or fully-grown mussels, from places where it has settled naturally. The seed which settles in January and February can be found on the rocks close to the surface of the water, but since the intertidal zone is very narrow in these waters with a small tidal range, the strips with mussel seed do not usually form a very rich source. Moreover, the mussel farmers have observed that small seed taken from the upper part of the intertidal zone often leads to very poor farming results. If the seed has settled in a place where feeding conditions are permanently sub-optimal, very slow growth may occur and a stunted type of seed is then the result. When such stunted seed, recognizable by the wide angle formed where the valves of its shell meet, is brought to a mussel park with excellent feeding conditions, it will usually fail to grow. This is a common experience in farming molluscs: stunted growth in early youth renders juvenile shellfish unfit for farming.

To scrape the seed from the rocks, which is done in various places along the coastline, including the riprap protecting the Diga Foranea and a further source formed by the large mooring buoys in the Roads of La Spezia, instruments called "raschiette", sharp steel blades of 25 to 30 cm broad to which a wooden shaft of 1.5 to 2 m length has been attached, are used. The seed which settled in winter is gathered several months later, from April to July. To this end one sails to the selected spot with the regular working boat [1] ("barca da lavoro"), equipped with an outboard motor (Fig. 65). One or two persons occupy the boat. Scraping seed from the rocks can only be done in calm and dry weather. If conditions are favourable, work starts early in the morning and finishes late in the afternoon. Usually one need not sail very far, but occasionally it is necessary to make a trip of several hours, for instance to Carrara, to find a productive place. A special permit is not required to collect mussel seed, but some sections of the naval base are out of bounds. The quantities of seed harvested per day vary a great deal; sometimes only 100 kg can be collected after 6 to 7 h work, but occasionally as much as 1 000 kg can be harvested per day. It is, however, not so much the weight of the seeds which is important for good farming, but rather their number.

Fig. 64. Headquarters of the mussel farmers of La Spezia, mainly used for storage of farming materials. (Photo Dr^{ssa} G. Nacini.)

Some fishermen sell the seed harvested to mussel farmers. They will receive L. 60/kg for the small seed collected in April, which may multiply its weight ten times during cultivation, but only L. 20 for the larger seed harvested in

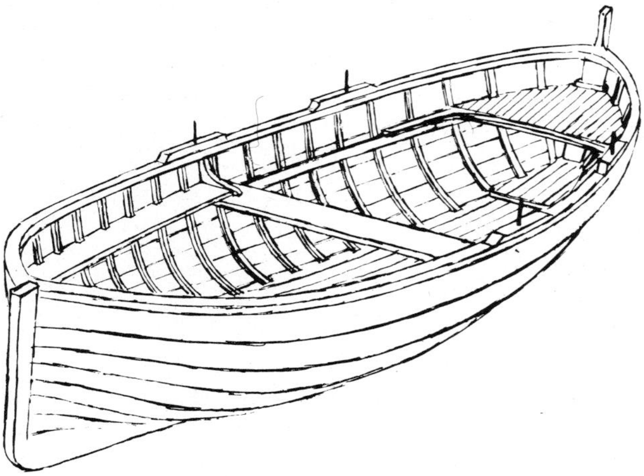

Fig. 65. Working boat of the type called "gozetto".

July. The cooperative "La Proletaria" participates in collecting seed from the rocks, but obtains only about 10% of the quantity required for farming this way, the other 90% being collected in its own parks with the aid of collector ropes. Therefore little time is devoted to such operations by the participants of the cooperative.

Larger mussels can sometimes be found on the poles of landing stages, on timber work or on ship wrecks, and it may pay to harvest this material and transfer it to the mussel parks. To collect such larger mussels from deeper sites a diving suit is occasionally used, but the normal procedure is to collect these mussels with the "brancone", a curved steel blade to which a net of 50 cm long and 30 cm diameter is fastened (Fig. 66), to catch the mussels which are scraped off. A wooden shaft, 3 to 4 m long, makes it possible to reach deeper sites. A piece of expanded polyester is usually attached to the "brancone", to prevent sinking if it is dropped in the sea. The cooperative "La Proletaria" does not participate in collecting larger mussels and does not buy such mussels from fishermen.

For this company the other way to produce the seed, viz. with the aid of collectors, is far more important. It covers 90% of their needs and occasionally so much is produced that part of it has to be sold to other farmers. The collector ropes, called "filimbindo" consist of three intertwined strings of synthetic material (polyethylene), together about 25 mm in diameter which have quite a shaggy appearance and which are therefore excellent collectors for mussel seed. The filimbindos are put in the sea in the month of December to collect seed in January and February. The material is quite strong and has a life of at least 4 years. The filimbindos are invariably hung horizontally

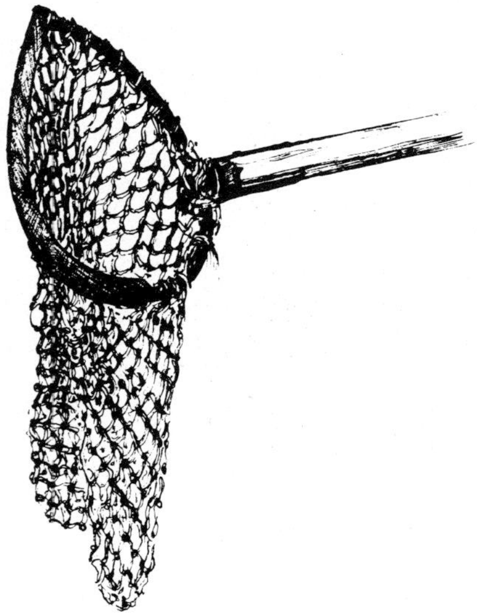

Fig. 66. The "brancone", a curved steel blade with net attached to it.

around the entire circumference of a mussel park ("vivai") [2], at a depth of about 50 cm, suspended by thinner ropes from all the obliquely-placed retaining poles. For a standard park measuring 40 × 25 m (total area 1 000 m^2) the filimbindo is approximately 135 m long. To keep the filimbindo well under water, the mussel farmer hangs one or two strings with mussels from the filimbindo itself in every section between two poles.

Often the mussel farmers apply a double filimbindo, again around the entire circumference of the park, but hung from separate supporting ropes so that it comes somewhat closer to the line of vertical poles, but about 20 cm deeper than the "normal" filimbindo. For a standard park of 1 000 m^2 with a double filimbindo, the collectors have a total length of about 270 m.

The seed will settle on the filimbindos in the first months of the year. Harvesting can start in April and can be continued till September. Usually the first crop is taken off by hand in April or May allowing the remaining mussels a better chance to grow, so that one can harvest a second crop later in summer. It has been stated above that 90% of the seed the cooperative "La Proletaria" needs for its farming is collected in this way.

B.2. Making strings of mussels from the seed

When the seed has been harvested, either from the rocks or from the filimbindos around the parks, it must be converted into strings to be hung verti-

Fig. 67. Demonstration of the old method to use a rope to hold clusters of mussels. (Photo P. Korringa.)

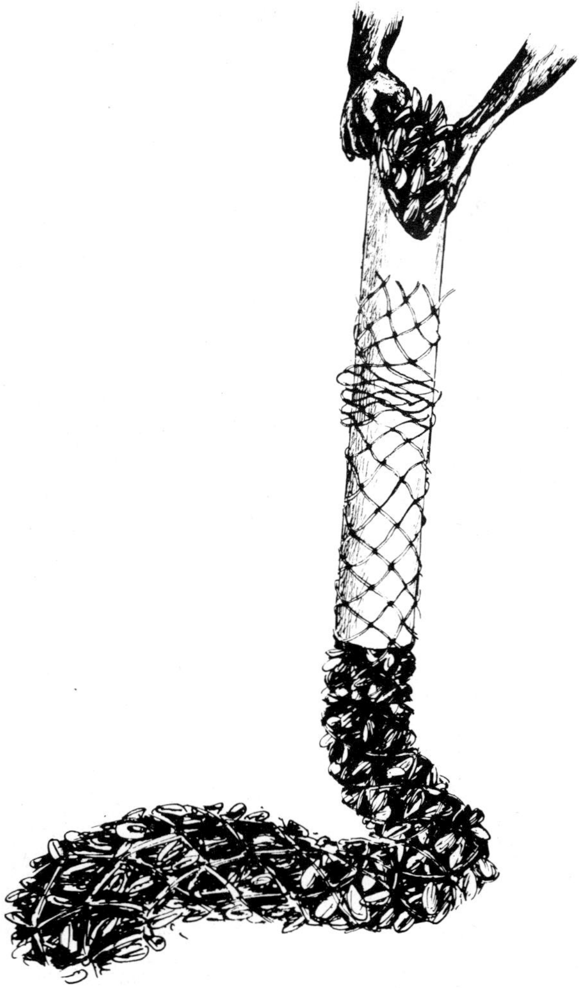

Fig. 68. Use of a plastic tube to fill the cylindrical nylon netting with mussels.

cally from the horizontal supporting ropes in the parks. In former years ropes made of sparto grass *(Lygium spartum)* were used for this purpose; this grass is locally called "erba palustre", and grows abundantly in several places along the Italian coast, for instance on the Isle of Palmaria. By twisting open such a rope (Fig. 67) one can clasp a cluster of mussel seed in between the strings so that it will not fall off when hung in the water. In due course the mussels will crawl about, attaching themselves where it is suitable with byssus threads. Though this system is still widely used, the cooperative "La Proletaria" has completely switched over to a more modern procedure, using stronger material. From the company R. di B. Viganò at Brianza

(Como), large rolls of a cylindrical netting made from strong synthetic threads 0.75 mm thick, called "Netlon" can be purchased. The "knots" of the meshes are sealed and there are three different gauges in use in the Italian mussel industry, viz. No. 14: stretched mesh 7 cm, diameter 9 cm; No. 12: stretched mesh 10.5 cm, diameter 11 cm; and No. 10: stretched mesh 14 cm, diameter 13 cm. Gauge No. 14 is used for mussel seed, No. 12 for half-grown mussels (2—5 cm shell-length), whereas No. 10 is only used occasionally to make strings of larger mussels. To put the mussels in the cylindrical netting one uses 1 m long cylindrical tubes which are open at both ends and made of polyvinyl chloride (PVC) (Fig. 68); for gauge No. 14 a tube with a diameter of 6 cm, for No. 12 one of 10 cm diameter and for No. 10 a tube of 13 cm diameter is used. The work proceeds as follows: the desired length of netting is cut from the roll with a small sickle (a curved steel blade 18 cm long with an undulating edge and a wooden handle) and one end is closed with a piece of synthetic string. Next one slips the piece of netting over the plastic tube. The tube is held vertically and filled by hand with clusters of mussels until they reach the top of the tube. Then the tube is lifted, allowing the mussels to sink into the netting which slides away from the tube under the weight of the mussels. The tube is then put down again and the filling procedure re-

Fig. 69. This basket, called "cesto", holds mussels which are for the time being insufficiently clustered to be put in netting. (Photo Dr[ssa] G. Nacini.)

sumed. The net is closed at its upper end with a synthetic string. The upper string should be longer than the lower one, for it should also serve for attaching the net to the supporting rope of the park. Thus, in a very short time, cylindrical nets of 4 to 8 m length filled with mussels are produced; it takes an experienced worker 6 to 8 min to prepare one string for hanging out in the park. These are the "pergolari", as the Italian mussel farmers call them. This work, making the pergolari, is always carried out in the open air. Sometimes it is done at the headquarters ashore, but about 70% is done on the working boats moored in the park. If so desired, one can keep the seed or the half-grown mussels dry overnight before making the pergolari. This system of filling the nets does not necessitate weighing the quantity of the mussels put in. If the seed is small one needs about 1 kg/m of No. 14 netting, and for half-grown mussels, it takes about 3 kg/m of No. 12 netting.

It is clear that the meshes of the netting are so big that they will not retain individual mussels. Therefore one can only make pergolari with well-clustered mussels, either as seed or half-grown. If a batch of mussels is not sufficiently clustered, it is placed for a couple of days in a special flat basket, called "cesto" (Fig. 69), hung horizontally from supporting ropes in the park, close to the surface of the water. Under such conditions healthy mussels will rapidly spin themselves together with byssus threads. Then they can be put in the nets.

B.3. Growth

The pergolari are hung from supporting ropes in the parks. If the nets have been filled with mussels ashore, they are transported by boat to the parks. Again, working boats with an outboard engine are often used, whereas in other cases one or more working boats are towed to the parks by a motor boat [3]. There the pergolari are fastened one by one to the horizontal supporting ropes at a distance of 60 to 70 cm from one another. The extra piece of string at their top serves this purpose. The poles of the park are spaced 5 m apart. From six to eight pergolari are hung from each rope connecting two poles ("ventia"). If the poles are connected crosswise with each other, nine pergolari are hung from such a cross ("crociera"). Often ropes are hung both on the inside and on the outside of a row of poles, making room for more pergolari. In any case they should be hung out in such a way that they do not interfere with each other. The number of pergolari that can be put in a park depends on the size and shape of the park. A park of 500 m^2 carries as a rule about 450 pergolari, and one of 1 000 m^2 about 715 pergolari. A smaller park has a relatively longer circumference, hence relatively more rope length for pergolari, but a larger park has the advantage that a smaller number of poles per 1 000 m^2 of park are needed.

The total length of netting per 1 000 m^2 of park depends on the length of the pergolari, which are from 4 to 5 m in parks in a shallow area leaving 60 to 80 cm between the muddy sea bed and the pergolari, and up to 8 m for parks in deeper water, leaving a distance of about 4 m between the sea bed and the pergolari. Even where depths would allow it, a length of mussel string is never greater than about 8 m, lest the load to be carried becomes too heavy.

The mussels in the filled nets are haphazardly positioned. Soon after the pergolari have been suspended in the park the mussels begin to crawl through the meshes of the netting, detaching and re-attaching their byssus threads in order to gain an optimal position for feeding, siphons outward. Gradually the netting material becomes buried inside the pergolari. By crawling out the mussels free themselves from the pressure of the netting caused by the weight of the whole pergolare. When the pergolare is lifted after some time one does not see the netting; it can be found in the middle, like the vertebral column in an eel. Some silt may accumulate around the netting and when the total weight of the mussels becomes too great, there is a growing risk that the whole load may slip from its support.

B.4. Thinning out

To prevent loss of mussels caused by slipping from the pergolari they should be distributed over a greater length of netting as soon as the weight of one pergolare reaches about 30 kg. To achieve this the pergolari are taken up and cut into sections of about 1 m length. The original pieces of netting in the centre are now easily pulled out and by the same procedure described above the mussels are put in new nets, now of larger mesh size and diameter, e.g., gauge 12. When the mussel seed has been hung out in summer, the mussels are usually put in new nets the following spring. When growth is exceptionally good it may be wise to change the netting again (now using gauge 10) in August. The same thing is done with mussels which for some reason or other have not been sold by September or October.

It is unlikely that the weight of the pergolari will exceed the critical level unnoticed, for the parks are inspected regularly. For farmers still using sparta ropes with their short life it is obvious that frequent inspection is desirable, but also when netlon is used to construct the pergolari one cannot leave them uninspected for a long time. Fouling organisms such as sponges, hydroids and ascidians may settle in such numbers on the mussels that their growth is impaired. If the farmer sees that such is the case, he will detach his pergolari from the suspending ropes, and hang them in the air for some time in order to kill these epibionts which cannot safely withdraw inside their shells. For this purpose he has constructed close to his park a "fusolo" [2], a simple rack with horizontal poles at about 2 m above the sea level (Fig. 70). He hangs the pergolari on the fusolo in such a way that they wholly are out of reach of the water and sprays them thoroughly with the aid of a motor pump to wash away the silt. He will usually do this in the early afternoon

Fig. 70. Placing the "pergolari" on the "fusolo" to destroy fouling organisms. (Photo Dr^{ssa} G. Nacini.)

and return the mussels to the water again the next morning. He has to take into account the weather prevailing that day. The sun may be so hot that his mussels are killed within their shells by overheating. If such is the case, he hangs thin sheets of burlap over his mussels. The fouling organisms will be killed by drying out, while the temperature of his mussels will not attain the danger level.

Putting the pergolari on the fusolo is a time-consuming operation; two men working together take about 1 h to detach 20 pergolari, bring them to the fusolo, and hang them out and spray them. Bringing them back again to their places in the park requires about half that time. One should reckon on a sojourn on the fusolo about once a month in the summer season, and once every other month in winter. The mussel farmers are convinced that such treatment also makes their mussels more resistant during transportation.

B.5. Harvesting of marketable mussels

When the mussels have attained the marketable size of 5 to 6 cm shell-length, as they usually do in the course of their second summer and when

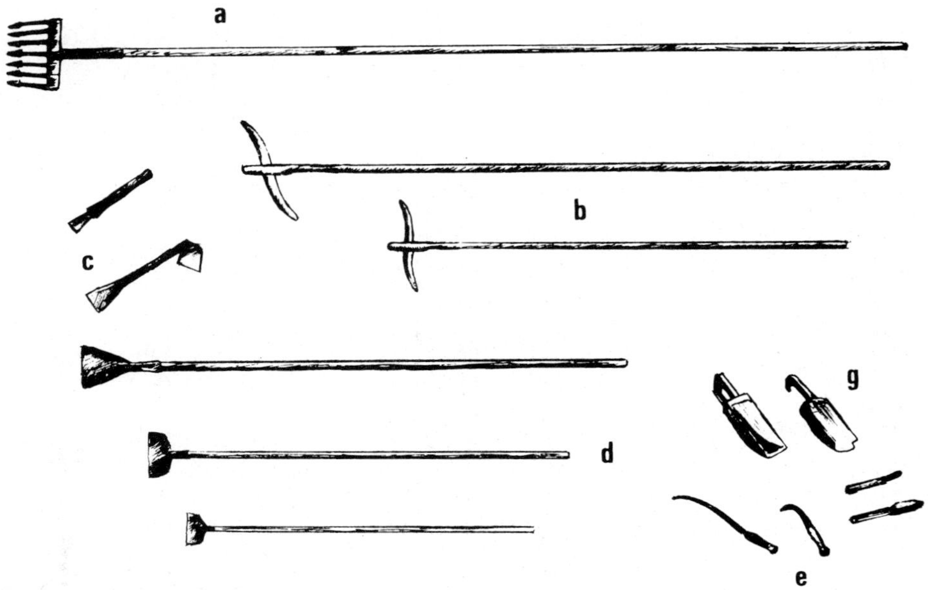

Fig. 71. Utensils used in mussel farming Italian-style: (a) "fiocine", fork to salvage lost "pergolari"; (b) "croci", to lift up "pergolari" in a park; (c) small scrapers for various uses; (d) "raschiette" to scrape mussel seed from rocks or poles; (e) sickles to cut nylon netting; (f) knives for cleaning mussels; all steel or with wooden handle; (g) bailers.

the edible tissues are in good condition, the farmer may proceed with the harvesting. This is quite a simple procedure in mussel farming Italian-style. The pergolari are cut loose and transferred by hand to the deck ("sana") of the working boat. The "croce", a wooden shaft with a curved piece of wood attached cross-wise at its end, is used to lift the pergolari (Fig. 72). With the croce, the heavy load of mussels is dragged towards and over the edge of the boat.

Often the mussels are cleaned on the deck of the boat. A blunt knife serves to eliminate as far as possible incrusting organisms such as barnacles and *Pomatoceros*, and a brush is used to get rid of softer fouling organisms and silt. The mussels are washed profusely at the park. Sometimes they are landed as complete pergolari, to be sold as such, and in other cases one cuts up the pergolari to clean and grade the mussels. The mussel farmer, his hand protected by a heavy glove, rubs his mussels over an iron grid to separate them. Other farmers use a sieve of synthetic netting for this purpose. It takes about 1 h to treat 100 kg of mussels this way, and per working day of 8 h one man can grade, clean and wash 500 kg of mussels. A jet of water from a motor pump assists him in this work. If so desired, and taking the weather and other things into account, one can proceed to wash and grade the mussels at the headquarters.

Fig. 72. Taking up a "pergolare" with full-grown mussels with the "croce". (Photo Dr^{ssa} G. Nacini.)

In harvesting marketable mussels two men can take up per day 320 pergolari of 4 to 5 m carrying 25 kg of mussels each, a total of 8 000 kg of mussels. It is clear that this does not include the grading and separating of the mussels, but only washing. The mussel farmers of La Spezia claim that more incrusting organisms such as "vescia" *(Pomatoceros triqueter)* grow on the shells of their mussels than in earlier years and they blame increasing pollution of the Roads of La Spezia for this. In the year 1969 they made an effort to mechanize the time-consuming cleaning of the mussels; a machine was developed which brushes the mussels with the aid of sand. The result was not encouraging; the mussels received so much damage that they died in transit to the market. The mussel farmer of La Spezia returned to the procedure of cleaning by hand.

Depending on the length of the pergolari, which is related to the local depth, one standard park of 1 000 m^2 can yield annually from 20 to 35 metric tons of marketable mussels.

When the park is situated in a section of the Roads of La Spezia influenced by sewers, an extra operation must be performed if a certificate of bacteriological purity for the mussels is required. The pergolari with marketable mussels must then be transported to a park situated in pure sea water (e.g., close to the Diga Foranea, or at the north coast of the Isle of Palmaria), where they should be suspended for several days. This means, of course, extra labour.

B.6. Shipment and marketing of mussels

Storage of marketable mussels is not practiced by the farmers of La Spezia. The quantities destined for immediate sale are harvested and cleaned. There are no basins or other storage facilities at the headquarters. Unsold batches are thus returned to the park.

There is no official close season for marketable mussels in Italy, but parallel to the quality and the demand for mussels there is a sharp seasonal fluctuation in the sales. The period of March to November should be considered as the main harvesting season, with a peak from June to August.

The mussels are sold to wholesalers either in bulk or packed in sacks or boxes. If they are sold in bulk, the pergolari are weighed (Fig. 73), after which they are loaded onto a truck. Though most of the mussels grown in La Spezia used to be sold in bulk, there is now a marked tendency to insist on

Fig. 73. Weighing strings of marketable mussels delivered by the farmers of La Spezia. (Photo Dr^{ssa} G. Nacini.)

the provision of mussels ready for the kitchen. Therefore, the percentage of separated mussels packed in boxes or sacks has increased sharply in recent years, though such treatment reduces the strength of the mussels, causing them to suffer more readily during transportation. In 1971 about 80% of the mussels were sold that way. After a period during which returnable plastic boxes, containing 25 kg of mussels were in use, now most of the mussels are sold packed in sacks of nylon netting, some also in burlap sacks, containing 25 kg, and some in simple non-returnable wooden boxes.

The mussels are sold to wholesalers at the headquarters of the farmer. They are shipped to the consumer by truck or by rail; in the warmer season, temperature-controlled trucks are employed. Only about 3% of the mussels are consumed in La Spezia itself, the main markets being Genova and Riviera di Ponente, Roma, Riviera di Levante, Milano, Torino, Versilia, Livorno and Puglia.

C. FARMING RISKS

C.1. Hydrographic conditions

Every winter parks are damaged by gales. The larger parks especially may suffer since the forces exerted at the corners of such parks are considerably greater than in the more soundly constructed smaller parks. Parks made of wooden poles, so easily attacked by shipworms, may suffer more readily than parks made of iron poles. As soon as damage occurs one should make an effort to salvage the mussels, since they perish very easily in the soft black mud on the sea bed. In shallow places the "croce" may serve for this purpose, but for deeper places, a kind of long shafted fork with barbed teeth, called "fiocine" is used. In general one should expect a loss of about 5% per year by gales. In less well-sheltered mussel regions of the Italian coast the annual losses by gales will be higher.

As stated above, some mussels may die in summer because of thermo-pollution in the vicinity of the power plant. In the area where this danger exists one should expect at least 1% loss.

C.2. Biological conditions

Though blooms of a noxious phytoplankton do occur occasionally from place to place on the Italian coast, and quite regularly in such fertile sites as the Lago Lucrino near Naples, the mussel farmers of La Spezia rarely suffer losses of mussels or unavoidable interruptions of their sales from this cause. 1948 was the last year in which such a noxious bloom was observed here.

C.3. Predators

Mussel farming in mid-water reduces the losses by predators to a very high degree, since starfish, crabs and the like cannot reach the mussels. In former years ravages caused by the Sparid fish called "orata" *(Aurata aurata)* were feared in this area, but for some reason, maybe because of the increased pollution of the water, these fish have not been noticed here for several years.

C.4. Parasites and diseases

No losses caused by parasites or diseases have been recorded in this area.

C.5. Competitors

If the farmers inspect their mussels regularly and make frequent use of the fusolo, they need not fear poor growth or losses due to fouling organisms such as ascidians, sponges, and hydroids. Incrusting organisms such as balanids and the "vescia" *(Pomatoceros triqueter)* are cursed by the farmers since they cause so much extra work to produce clean mussels, although they do not actually kill the mussels. There is some concern about a mass development of the seaweed *Codium* in recent years, which may lead to serious fouling. It is surely possible to control this competitor in some way, but the Italian Government does not carry out biological research to assist the mussel farmers in dealing with such problems.

Indirectly the shipworm *Teredo* is a serious enemy of the mussel farmer, especially where he is still using wooden poles to construct his parks. Such poles have to be replaced more frequently in areas where shipworms abound. Poles of poplar or pine have a very short life here. Poles made of the wood of the sweet chestnut *(Castanea vesca)* are much better; it is said that the tannin content of the bark keeps away the shipworm. The mussel farmers claim that sea urchins may gnaw at the bark close to the sea bed and in doing so allow further attack by the shipworms.

D. ECONOMIC ASPECTS

Dr[ssa] Giuseppina Nacini, wife of Dr M. Bernhard, director of the Fiaschery Laboratory, has made a detailed study of the economy of the mussel industry at La Spezia. She has kindly offered her comprehensive data for perusal, and this forms the basic material for the considerations below.

D.1. Rents

It has been mentioned above that the Italian Government promotes exploitation of mussel parks by cooperatives by granting them very favourable leasing terms, in this case an annual rent of L. 5 000.

D.2. Inventory items

Among the inventory items the fleet ranks first. The cooperative "La Proletaria" operates eight motor boats, 18 working boats and two smaller boats called "gozzetto". There are moreover five outboard engines in stock which can be mounted on the working boats when there is no motor boat available. A motor boat of the type used by the mussel farmers of La Spezia can be purchased for about L. 700 000 to which should be added L. 800 000 for a 12 to 15 h.p. engine. It may have a life up to 25 years, but after some 15 years considerable repair work can be expected. The engine too may last 25 years, but when new and better types of engines are on the market one may not be able to resist the temptation to install them. The maintenance costs of such a boat will be about L. 50 000 per year including painting twice a year.

The working boats, constructed of wood, cost about L. 700 000 each. If properly maintained they will last at least 20 years, and the annual maintenance costs are in the same order of magnitude as those for the motor boats. A 7 h.p. outboard engine can be purchased for L. 400 000 and has a life of about 4 years.

Per 1 000 m^2 of park area one should budget for an expense of L. 54 000 for outboard engine fuel (average use: 2 h per day, 25 days per month) and L. 10 000 per year for diesel oil for the motor boat. The motor pumps require fuel costing L. 10 000 per 1 000 m^2 of park area per year.

The number of motor boats in use by the cooperative "La Proletaria" is greater than in earlier years. There was a time when only three motor boats were operated, but it required detailed planning and a great amount of cooperation to do all the work properly and efficiently with so few motor boats on the 27 000 m^2 of park. To prevent idleness and to gain some more freedom in programming the work of the individual participants of the cooperative, the number of motor boats was gradually increased to eight, more than really necessary for the sake of economy.

Other inventory items are the motor pumps used on board for washing the mussels that are hung in the air on the fusolo and those ready for marketing. These pumps cost L. 60 000 and have a life of 4 years. Ashore one needs the following items: a derrick, placed on the wharf, to hoist up boats and engines, costing L. 250 000, with an estimated life of 25 years; welding apparatus to make iron poles of the desired length, costing about L. 200 000 with a life of 20 years; a pump for washing the mussels, costing L. 700 000 with a life of 12 years; a weighing machine with a capacity of 100 kg, costing L. 300 000, life 4 years; a non-automatic balance, cost L. 50 000, lasting 8 years; a Jacob's ladder (L. 500 000), to load the sacks of mussels onto the trucks, life 15 years; and a 5 m long barrow on two pneumatic tyres for transporting poles to the boat.

One also needs a landing stage, and a shed built on poles to store various materials. The shed of the cooperative "La Proletaria", is made of wood and covered with a roof of corrugated iron, and measures 5 × 3.5 m. Finally, mention should be made of the office for the administration, equipped with the usual furniture and apparatus, including a telephone and a calculating machine.

D.3. Materials

Purchase of wooden or iron poles is a recurring item on the budget of the Italian mussel farmer. To construct a standard park of 1 000 m^2 one needs 116 poles, differing in length from 13 to 15 m for sites 8 to 9 m deep, from 16 to 18 m for depths of 11 to 12 m. One pole of chestnut costs about L. 6 000; they last only 2 years. The cooperative under consideration nowadays uses iron poles almost exclusively (92%) which at L. 14 000 each cost more than double the price of wooden poles but will last longer (5 years) and hence require less labour for replacement. 90% of the poles used by the cooperative are pipes of plain iron (76 mm diameter, wall 3.75 mm thick) costing L. 1 000/m, but they experiment with galvanized iron pipes costing L. 1 200/m, and with pipes coated with a mixture of bitumen and glass wool (L. 1 200/m) which are said to last for at least 7 years.

Where parks are constructed in deeper places the cost of poles per 1 000 m^2 of park area is higher, but in such parks the length of the pergolari can be greater than in shallow sites, which compensates to some extent for the higher cost of poles.

For the construction of the parks polyethylene ropes, 14 mm thick are needed. For a standard park of 1 000 m^2 one should budget for 1 900 m of rope, which costs L. 160/m, and lasts 3 to 4 years. In addition, one needs a thicker rope for the filimbindo (275 m) and a certain length of 8 mm polyethylene rope, firstly to tie the poles together at the circumference of the park, and secondly to suspend the filimbindo from the oblique poles.

A further item is the "netlon" netting to make the pergolari. A park of 1 000 m^2 carries some 720 pergolari 4 to 8 m long. First the seed is put in the netting No. 14, to be transferred a couple of months later to No. 12 netting, and often it is necessary in the same year to transfer the mussels to No. 10 netting. Therefore one should expect to purchase 8 to 16 km of "netlon" (L. 6.50/m) per park of 1 000 m^2 per year.

The cost of packing material (boxes or sacks) is estimated at L. 5/kg of mussels, whereas a modest amount of money is required for electricity ashore for illumination, for welding, and for operating the pumps and the Jacob's ladder.

Not counting labour, Dr.ssa Nacini concludes that the costs per 1 000 m^2 park area are L. 907 500 for cultivation, and L. 350 000 for marketing the mussels.

D.4. Labour

From the description of the way mussels are farmed in the area under consideration, it can be seen that it requires a considerable amount of manual labour. Apart from the use of engines in the boats and pumps, very little has changed in the labour pattern in the course of this century. The main jobs to be done are welding poles together to the desired length, constructing parks with poles and ropes occasionally, replacing poles and ropes, collecting mussel seed, making pergolari from the seed, hanging the pergolari in the parks, transferring half-grown mussels into a greater length of netting, putting the pergolari repeatedly on the fusolo, transferring pergolari from polluted parks to clean parks, harvesting the pergolari carrying marketable mussels, cleaning, grading, washing and packing the mussels.

It is difficult to estimate the exact number of working hours required for each of these items, but according to Drssa Nacini one experienced mussel farmer can deal with a park of 1 000 m^2. Considering that the cooperative "La Proletaria", which includes 18 participant-proprietors, exploits 27 000 m^2 of parks and employs in addition four hired hands all the year round, and during great pressure of work about 40 extra mandays, the assumption that one man can farm a park of 1 000 m^2 cannot be far from the truth.

D.5. Sales

The price for which the marketable mussels are sold to the wholesalers is L. 115/kg for mussels packed in sacks and boxes, and L. 80 for mussels in bulk, i.e., the complete pergolari. The retail price is about L. 250. Total production of marketable mussels fluctuates somewhat from year to year. The cooperative "La Proletaria" sold about 570 metric tons of mussels in the year 1968 and 450 tons in 1970. During the last few years they could not sell all the mussels produced, partly because of severe competition from the import of Spanish mussels which are of larger size (8 to 9 cm against 6 to 7 cm for the mussels of La Spezia) and often have a better meat content, especially in the winter months. Therefore the Spanish mussels can be sold to the wholesalers for L. 200 to 250/kg and fetch L. 400 retail.

D.6. Losses

One should expect 5% loss from damage to the parks caused by gales and

1% loss from thermo-pollution in a part of the Roads of La Spezia. Theft is of rare occurrence since paid watchmen guard the area.

D.7. Personnel

The participants of the cooperative "La Proletaria" carry out most of the work themselves, both at sea and ashore. They hire four extra hands on a full-time basis at a cost of L. 175 000 per month, taxes etc. included, and pay about L. 200 000 per year (L. 5 000 per day of 8 h) for extra seasonal labour. The cooperative has also to pay the watchmen L. 58 000 per month.

D.8. Further developments

Production costs have gone up in recent years, not only because of the general increase in the price of labour, but more specifically because of the increasing pollution of the Roads of La Spezia which forces the farmers to confine their work more and more to parks further away from their headquarters, where the water is usually deeper, requiring longer poles for construction of the parks. It is claimed, moreover, that the omni-presence of incrusting organisms requires more work to clean the mussels than in former years.

Though the production price goes up, it is difficult to set a higher price on the market because of competition from Spanish mussels of better quality, especially on the markets of Torino, Milano, Genova and Roma.

There is a shortage of parks which produce mussels of top quality. Expansion of the production is not considered for the time being since this involves the hiring of expensive labour, and the cooperative is nowadays faced with the difficulty of selling all the mussels produced; about 10% have had to be brought back to the parks in recent years.

E. GOVERNMENTAL ASSISTANCE

The Italian Government assists the mussel industry in various ways, e.g., by leasing the parks to them at a low rent, especially to cooperatives, and by offering the mussel farmers loans at a very low rate of interest. Furthermore there is an official bacteriological standard of control on the basis of which the "Ripartizione Igiene e Sanita" of the "Comune della Spezia" issues certificates of cleanliness, enhancing public confidence in the safety of the local product.

Moreover there is the "Consorzio Mitilicoltori", officially still called "Consorzio Ostricoltori" as it was prior to 1919 when oyster culture was still practised here, which intervenes officially in cases of emergency such as occurred during the incidental oil pollution in 1968.

The Italian Government does not provide for biological research to assist the mussel farmers.

F. GENERAL IMPRESSION

Mussel farming as carried out in the Roads of La Spezia is an excellent example of rational exploitation of a body of sea water rich in phytoplankton. The ideal shelter offered by the western peninsula makes it possible to construct mussel parks at comparatively low cost. The percentage loss of mussels during the farming procedure is so low that it is hardly worth mentioning.

Reasons for concern are: the very low degree of mechanization; the large number of man-days required to produce one ton of marketable mussels; the difficulty of attracting young farmers to take over from the older ones in the cooperative and therefore a steadily increasing age among the Spezian farmers; the increasing pollution of the water leading to greater expense, because parks have to be constructed in deeper sites further away from the headquarters, and mussels have to be taken to pure water before they are ready for market, and, so it is assumed, because there is a higher incidence of incrusting organisms on the shells of the mussels. The net income per participant in the cooperative is too modest to attract the younger generation.

When the increasing competition from imported Spanish mussels is added to all this it must be concluded that the future of mussel farming in La Spezia is not very bright.

G. TECHNICAL DETAILS

G.1. Working boat

The working boat or "barca da lavoro" is constructed locally from oak 2 cm thick and measures 5.80 m in length, 2.20 m in beam and 0.80 m in height. Aft it has a "sana", a kind of working deck on which various operations can be performed, allowing the worker to maintain an upright position. The remaining three-quarters of the boat is open with one or two benches for an oarsman. A boom with a hand-winch can be placed on the working boat when poles have to be hoisted in the park during replacement or after damage caused by gales.

G.2. The mussel park or "vivai"

The mussel parks, consisting of a network of poles connected by horizontal ropes, are of rectangular shape but different dimensions. They range from 400 to 2 000 m^2 and are composed of square sections measuring 5 $\times$ 5 m,

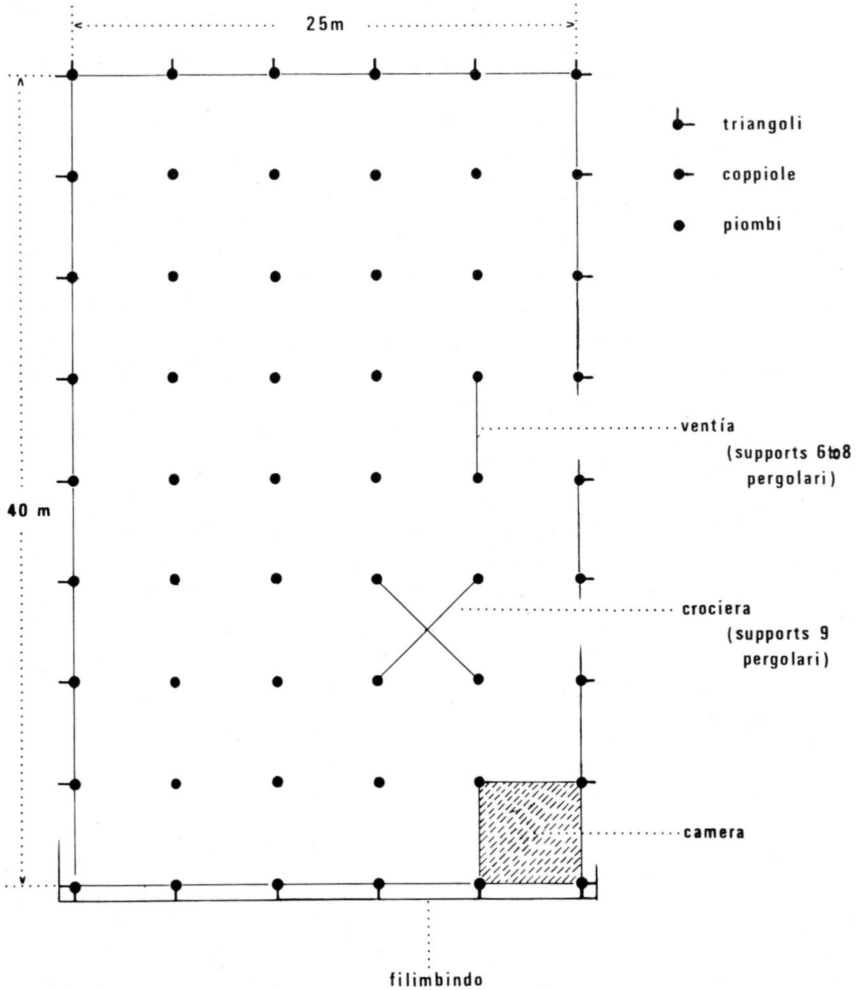

Fig. 74. Ground plan of a mussel park of 1 000 m^2.

called "camera", with poles at the corners (Figs 74 and 75). In between the parks one leaves wide corridors (30 to 50 m) for manoeuvering the boats.

To build a vivai one first prepares poles of the desired lenght, for shallow parks 13—15 m long, for the deeper parks 16—18 m long. If wooden poles are used one has to combine two chestnut saplings to make one pole of the desired length. The iron pipes (76 mm diameter with walls 3.75 mm thick) are commercially available in lengths of 6 m. Therefore several have to be welded together to attain the desired length. A piece of tubing fitting tightly into the pipes gives more strength to the connections. At both ends, the pipes are closed by welding on lids, so that the poles will remain air-filled when placed in the park. About 2 m from the bottom end of each pole a

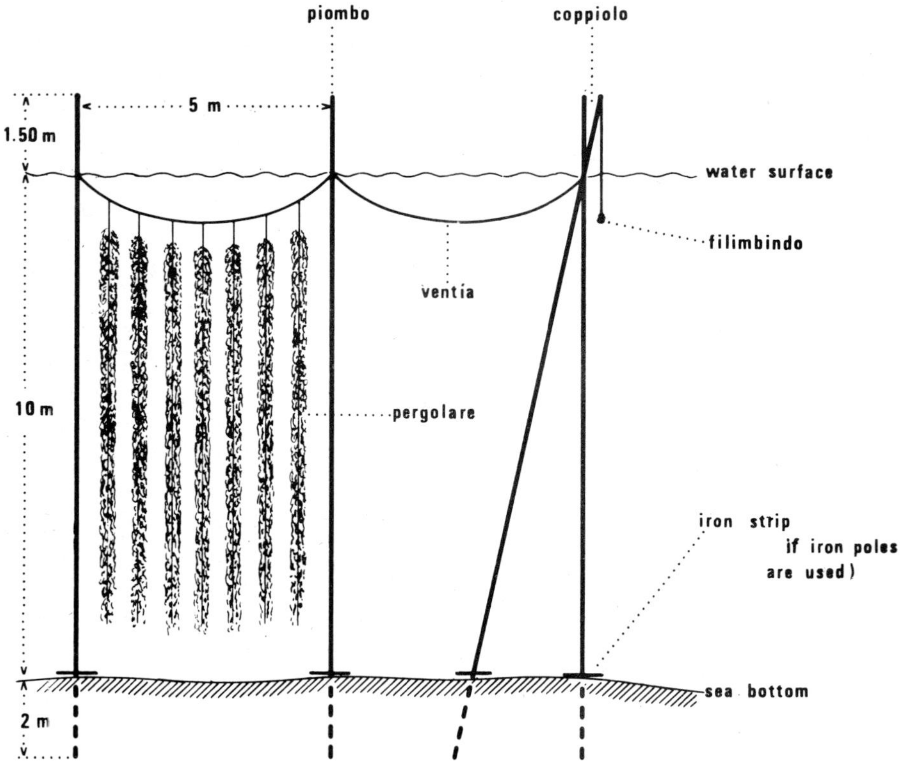

Fig. 75. Mussel park of La Spezia, seen in section.

cross-piece of iron of about 1 m length is welded on, flat side upwards.

The poles are transported on the working boat, tugged by the motor boat, to the sites where the park is to be constructed. One loads either six or seven poles length-wise on a boat, or one uses two working boats with about 20 poles lying across them. At the site the poles have to be placed in neat rows, 5 m apart, so that a system of 5 m squares, the camera, is formed. Such a park can for example measure 40 × 25 m, consisting of 40 cameras covering 1 000 m^2. When the correct place for a pole has been established it is driven about 0.5 m into the soft sea bed with a maul. A piece of rope is then attached to its upper end in such a way that a large noose is formed. Then the worker puts one foot in the noose and transfers his body weight to it, causing the pole to sink further until the iron cross-piece rests on the sea bed. The pole is then sufficiently deep — 2 m beneath the sea bed surface. Vertical poles are called "piombi". At the circumference of the park, an extra pole is erected near each vertical pole, and placed obliquely into the sea bed some distance inward from the perimeter of the park, at an angle of about 30°.

The two poles are tied together with polyethylene rope where they meet, just below water level. The pairs of poles at the circumference of the park are called "coppiole". At the four corners of the park further strengthening is necessary, since much of the weight of the mussels is indirectly transmitted to these corners. Therefore an extra oblique pole is placed there. Such a corner construction of three or four poles is called a "triangolo".

When the main structure of the park has been completed the poles are connected with horizontal ropes of 14 mm polyethylene just at the water level. These horizontal ropes will serve as supports for the pergolari carrying the mussels. Poles can be connected with one another over the shortest distance (5 m) with a piece of rope (called "ventía") which supports six to eight pergolari, or across the diagonals of the camera, then called "crociera", supporting nine pergolari. Usually ropes are connected at both sides of the rows of poles to make extra supports for pergolari.

In the month of December a 25 mm thick rope, the "filimbindo", is hung all around the circumference of the park, suspended by thinner ropes from the emerging top ends of the oblique poles. The filimbindo serves as collector of mussel seed. Often a double filimbindo is constructed, as described in the section dealing with procuring the seed. Depending on the material chosen for the construction of the park, one has to replace poles and ropes when necessary in order to prevent pergolari laden with mussels falling onto the muddy sea bed.

G.3. The motor boats

The motor boats (Fig. 76) used by the cooperative "La Proletaria" are constructed from 2 cm thick oak and measure 6.5 to 7 m in length, 2.30 m in beam and 0.90 m in height. A 12—15 h.p. engine (brand "Arona") is placed in the centre of the boat, protected against the weather by a roof; in front of it is a deck. The superstructure is made of pine.

Fig. 76. Motor boat as used in La Spezia.

H. BIBLIOGRAPHY

Berner, L., 1935. La Réproduction des Moules Comestibles (*Mytilus edulis* L. et *Mytilus galloprovincialis* Lmk.) et leur Répartition Géographique. Bull. Inst. Océanogr. Monaco, No. 680, 8 pp.

Berner, L., 1939. Nutrition et Infestation de la Moule de Provence (*Mytilus galloprovincialis* Lmk.) Bull. Inst. Océanogr. Monaco, No. 778, 27 pp.

Bluhm, P., 1955. Ein Vergleich zwischen der deutschen Miesmuschel (*Mytilus edulis*) und der italienischen (*Mytilus galloprovincialis*) an der Hand metrischer Merkmale. Arch. Fischereiwiss., 6: 334—342.

Carazzi, D., 1898. Ostricoltura e Mitilicoltura. Hoepli, Milano, 202 pp.

Cerruti, A., 1941. *Mytilhydra polimantii*, n.gen., n. spec., idroide vivente sul mantello dei mitili. Riv. Biol. Ital., 32 (20): 1—20.

Cerruti, A., 1948. Appunti sulla molluschicoltura in Italia. Boll. Pesca, 5—6: 7 pp.

Coppini, R., 1968. L'élevage des mollusques en Italie et ses aspects économiques et sanitaires. FAO Gen. Fish. Counc. Mediterr. Stud. Rev., Roma, 5: 473—477.

Garlanda, A., 1958. La Mytilicoltura nel Golfo di La Spezia. Ann. Fac. Med. Vet. Torino, 8: 52 pp.

Korringa, P. and Postma, H., 1957. Investigations into the fertility of the Gulf of Naples and adjacent salt water lakes, with special reference to shellfish cultivation. Pubbl. Stn. Zool. Napoli, 29: 229—284.

Mascellaro, A., 1961. Depurazione dei molluschi eduli. Thalassica Jonica, 4: 195—198.

Meyer-Waarden, P.F. and Mann, H., 1954. Untersuchungen über die Bestände von *Mytilus galloprovincialis* an der italienischen Küste auf ihren Befall mit *Mytilicola intestinalis (Copepoda parasitica)*. Boll. Pesca Piscicolt. Idrobiol., 8 N.S. (2): 5—24.

Nassogne, A., 1972. Etudes Préliminaires sur le Rôle du Zooplancton dans la Constitution et le Transfert de la Matière Organique au Sein de la Chaîne Alimentaire Marine en Mer Ligure. Assoc. Comm. Europ. Energ. Atom. Euratom/Com. Naz. Energ. Nucl. CNEN Eur., 4823 f, 237 pp.

Parenzan, P., 1958. Indagini esequite per il disciplinamento giridico, tecnico e sanitario della molluschicoltura e del commercio dei molluschi eduli in Italia. Thalassica Jonica, 1: 5—60.

Renzoni, A., 1960. Osservazioni sulle gonadi di mitilo (*Mytilus galloprovincialis* Lam.). 1. Variazioni stagionali del glicogeno. Pubbl. Stn Zool. Napoli, 32: 9—16.

Renzoni, A., 1961. Variazioni istologiche stagionale della gonadi di *Mytilus galloprovincialis* Lamk. in rapporto al ciclo riproduttivo. Rev. Biol., 54: 45—59.

Renzoni, A., 1961. Basi Biologiche e Prospettive di Potenziamento della Mitilicoltura nel Golfo di Napoli. Mem. No. 4, Dir. Pesca Marit. Minist. Mar. Mercant., 43 pp.

Renzoni, A., 1963. Ricerche ecologiche ed idrobiologiche su *Mytilus galloprovincialis* Lam. nel Golfo di Napoli. Boll. Pesca Piscicolt. Idrobiol., 18 N.S. (2) 187—238.

Strusi, A., 1964. Su alcuni caratteri chimici dei mitili (*Mytilus galloprovincialis* Lamarck) coltivato nel Mar Piccolo e nel Mar Grande. Boll. Pesca, Piscicolt. Idrobiol., 19 N.S. (2): 199—218.

Chapter 6

MUSSEL FARMING ON THE COAST OF ISTRA, YUGOSLAVIA

Example
Uzgajalište školjaka (Shellfish farm), Pomer, Istra. Leaders: Pavle Orešnjak and Ivan Poropat at Pomer, Ivan Janko at Pula.
Period of observation: 1972.

A. BACKGROUND

A.1. General principles

Mussel farming Yugoslavian-style resembles in many respects the procedure used in other Mediterranean areas. Since the tidal range is limited, and since the sea bed is often rather muddy because of the low current velocities prevailing under such hydrographic conditions, the best type of farming is rack cultivation, in which the mussels are kept well above the sea bed, but always under water. Sheltered conditions are essential and since at least some of the mussels are eaten raw, polluted waters should be avoided, at least for the last phase of cultivation. Mussel seed is neither gathered from the natural beds nor scraped from the rocks, but is procured with the aid of plaited vegetable collectors, stretched horizontally just under the surface of the water. When the mussel seed has been harvested from the collectors it is put into sections of cylindrical nylon netting, to be hung out in the parks. The fastest growth rate may be observed in slightly polluted water, but such mussels must be transferred to bacteriologically clean water some time prior to marketing.

A.2. Biology of the species

The biology of the mussels farmed in the Mediterranean has been described in the chapter dealing with mussel farming at La Spezia.

A.3. Geographical situation

The mussel farm under consideration operates both in the Vela Draga Bay and in Pomer Bay (see Charts). Vela Draga Bay is to be found due north of the harbour of the city of Pula at the further end of an inlet penetrating about 4 km into the Istra peninsula, at a latitude of 44° 53′ N. Pomer Bay is a well-sheltered inlet on the southern tip of the Istra peninsula. Pomer Bay has a rather irregular shape, with a maximum length and breadth of approxi-

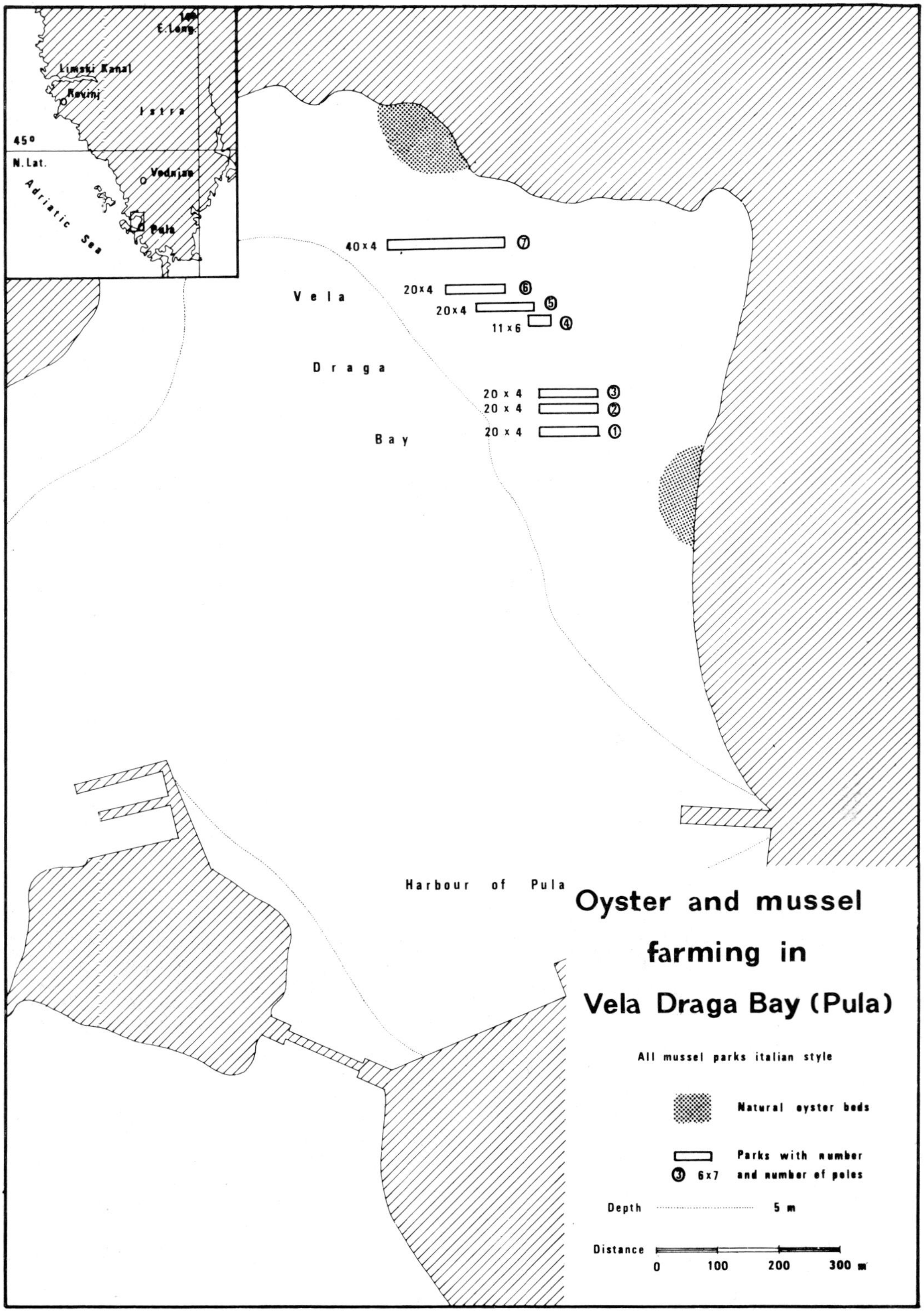

14°
E. Long.
Limski Kanal
Rovinj
Istra
45°
N. Lat.
Vodnjan
Adriatic Sea
Pula
40x4
⑦
20x4
⑥
20x4
⑤
11x6
④
20 x 4
③
20 x 4
②
20 x 4
①
Vela
Draga
Bay
Harbour of Pula
Oyster and mussel farming in Vela Draga Bay (Pula)
All mussel parks italian style
Natural oyster beds
Parks with number and number of poles
③ 6x7
Depth 5 m
Distance
0 100 200 300 m

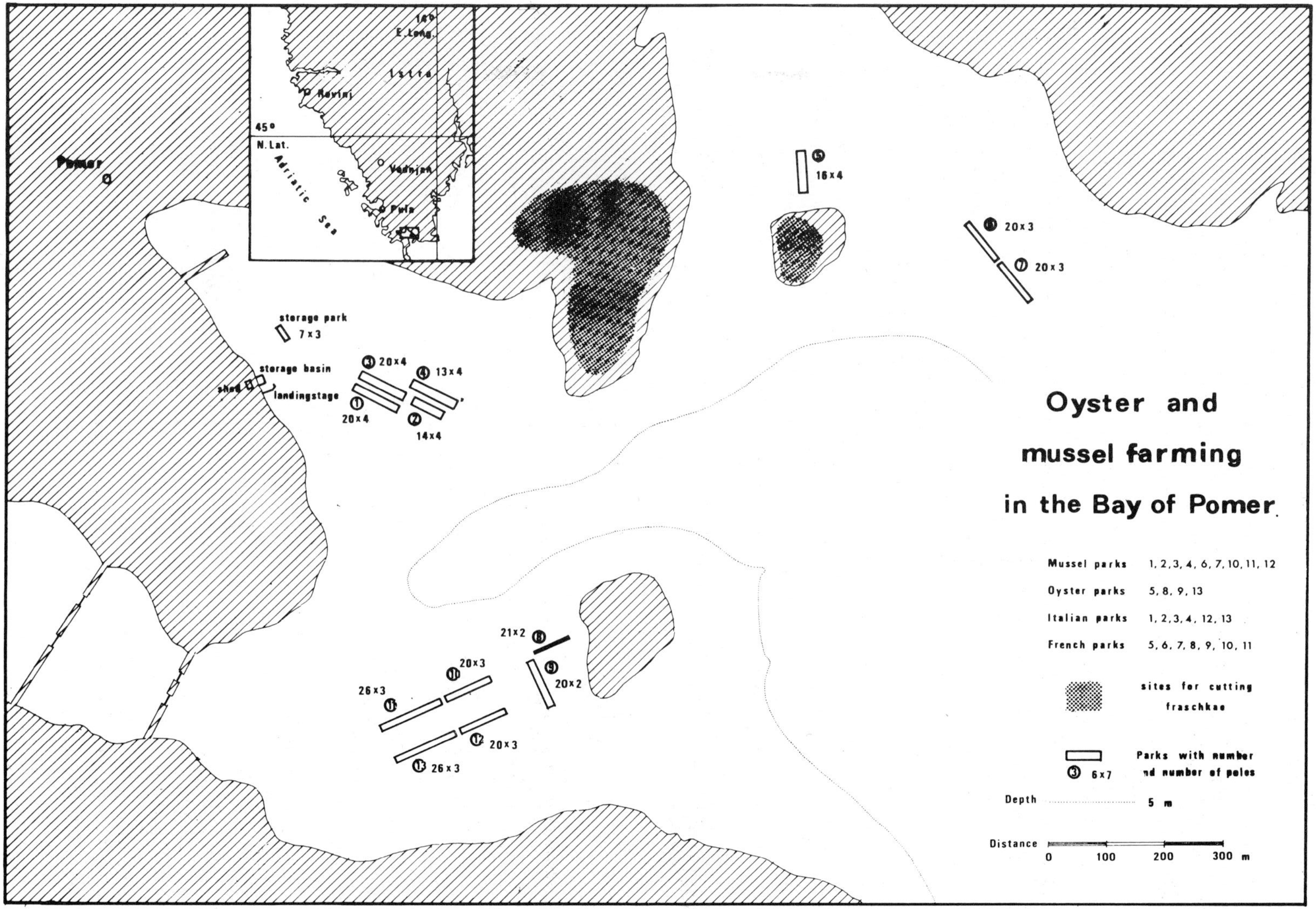

Oyster and
mussel farming
in the Bay of Pomer
Mussel parks 1, 2, 3, 4, 6, 7, 10, 11, 12
Oyster parks 5, 8, 9, 13
Italian parks 1, 2, 3, 4, 12, 13
French parks 5, 6, 7, 8, 9, 10, 11
sites for cutting fraschkae
Parks with number nd number of poles
6x7
Depth 5 m
Distance
0
100
200
300 m
Pomer
14°
E.Long.
Istra
Rovinj
45°
N.Lat.
Adriatic Sea
Vodnjan
Pula
storage park
7x3
storage basin
shed
landingstage
20x4
20x4
13x4
14x4
16x4
20x3
20x3
21x2
20x2
20x3
26x3
20x3
26x3

mately 3 km. The headquarters of the enterprise, where the shed and storage basin are to be found, is situated due south of the town of Pomer at the exact geographical position 44° 49′ 22″ N, 13° 54′ 04″ E. There is also a shed on the shore of Vela Draga Bay at 44° 53′ 01″ N and 13° 51′ 59″ E.

A.4. Hydrographic pattern

(a) Vela Draga Bay. Shellfish parks for farming oysters and mussels are laid out in the shallower parts of Vela Draga Bay, due north of the harbour of Pula. The depth ranges here from 3 to 5 m. The tidal range varies from 70—80 cm at neaps, to 115—130 cm at spring tide, quite a large range under Mediterranean conditions, due to the elongated shape of the Adriatic. The tidal currents are not very strong, and are largely governed by the wind. In winter, water temperatures drop below the 10°C level, but will only rarely drop below 8°C when the cold Bora wind occasionally blows for many days in succession. In summer, maximum water temperatures attain 26 or 27°C. The difference in water temperature between the surface layer and at 2 m depth is usually less than 1°C.

Since several springs and one rivulet flow into this part of the bay, the salinity may drop in the surface layers in periods of heavy rainfall. At 2 m depth the salinity fluctuates in the course of the year between 31 and 38‰ owing to variations in the output of the Po river, but in the surface layer salinities between 10 and 15‰ have been measured occasionally.

Oxygen measurements have shown that the water of the bay is usually saturated, but that somewhat lower values may be observed in late summer or early autumn.

(b) Pomer Bay. In Pomer Bay too, the shellfish parks are situated in rather shallow areas with a depth of 2.5 to 5 m. The tides show a range from 70 to 80 cm only, and current velocities are therefore low. Minimum water temperature in winter may be 5°C when the Bora blows, whereas in summer maxima up to 26°C have been measured. The influx of fresh water is very limited and hence one rarely measures salinities lower than 30‰ here. The oxygen content of the water varies between 70 and 120% of the saturation value.

A.5. Legal aspects

The Pomer shellfish farm called "Uzgajalište školjaka" is a section of the State Agricultural Cooperative "Poljoprivredna zadruga '1 maj' " which has its headquarters at Vodnjan, located 10 km north of the city of Pula, and which covers an area of about 40 km^2. Agriculture, including the production of fruit, wine and olives, and animal husbandry, constitute the main activities of this cooperative, which operates 25 shops for the distribution of its products. The Pomer shellfish farm is a section which functions with its own budget and therefore has its own administration. The director of the cooperative, agricultural engineer Miroslav Kapetanović, directs the shellfish farm too, and shows a keen interest in this enterprise, entrusting the managers, Ivan Janko at Pula, and Pavle Orešnjak and Ivan Poropat at Pomer, with the daily management.

The accounts of the shellfish section have to be audited by the accountants of the cooperative and these economic experts decide what should be done if the shellfish section makes a profit or suffers a loss. In the former case investment can be authorized to increase the production or to improve the production methods, and/or to pay the participants a share of the profit according to certain rules; in the latter case it has to be decided whether it is advisable to continue the venture or not. If the advice is positive, the shellfish enterprise can be subsidized from other sections which do make a profit. In the case of shellfish farming it is taken into consideration that its products are an attraction in the tourist industry and so some financial support is fully justified.

Since the shellfish farm is a State enterprise it encounters little difficulty in obtaining permission from the Town Council either to construct a park for growing oysters or mussels (provided that navigation is not hindered), to cut brush wood for collectors from State property, or to select sites for placing collectors. No rents need to be paid for the parks.

Some surveillance is necessary and therefore the shellfish farms at both Pula and Pomer Bay have a watchman to prevent damage or theft by tourists or poachers.

B. SEQUENCE OF OPERATIONS

B.1. Preparing the collectors

The Yugoslavian mussel farmers collect all the seed they need with special collectors made from the leaves of Sparta grass *(Lygium spartum)* imported from Italy. Ropes made of this grass, locally called "zukva", are purchased from a wholesaler in Trieste, who acquires it in turn from the production centres in southern Italy. When plaited, the rope is some 20 mm in diameter, and three of these ropes are interwoven into a kind of chain pattern. Such a chain has a diameter of about 10 cm and offers the mussel larvae a considerable area to settle on. Starting with the imported ropes one worker can make 200 to 300 m of chain-rope per day. The work is done in the yard of the shed at Pula.

B.2. The mussel parks

For collecting mussel seed, this company makes exclusive use of its parks in Vela Draga Bay. All these parks are Italian in style, as described in the chapter on mussel farming in Italy. The company now has seven parks in use in this area with a total area of 8 530 m^2. For growth and fattening of the mussels, the same parks in Vela Draga Bay are used, but in addition, this phase of the cultivation takes place on the parks numbered 1, 2, 3 and 4 (Italian-style), and 6, 7, 10, 11 and 12 (French-style) in Pomer Bay, with a total area of 8 270 m^2. The latter parks serve also for the cleansing of marketable mussels.

B.3. Bringing out the chain-rope collectors

Late in February or early in March the chain-ropes are brought to the parks in Vela Draga Bay. They are stretched length-wise over the parks, three in each interspace between the rows of poles, 3 to 3.5 m apart from each other. The thick nylon cross ropes connecting the poles carry the chain-ropes so that the latter cannot sag appreciably, but remain 10 to 30 cm under the surface of the water (Fig. 77). It should be clearly understood that the same parks are used continuously for growth and fattening of older mussels which are hung from the thick nylon or coco-fibre ropes stretched length-wise and cross-wise between the poles. The Italian system of putting the collector ropes for mussel seed around the park ("filimbindo") is seldom used here, since the park perimeter is needed for oyster spat collecting and hanging out ropes carrying oysters and mussels.

B.4. Maintenance of the chain-ropes on the park

The mussel seed settled in late winter or early spring grows rapidly and may attain a shell-length of 2.0—2.5 cm at the end of the summer season. When settling and growth are satisfactory the chain-ropes coated with mussel seed will in due course reach a diameter of up to 40 cm, and will be found to carry some 10 kg of mussels per m length. The rope is therefore quite heavy

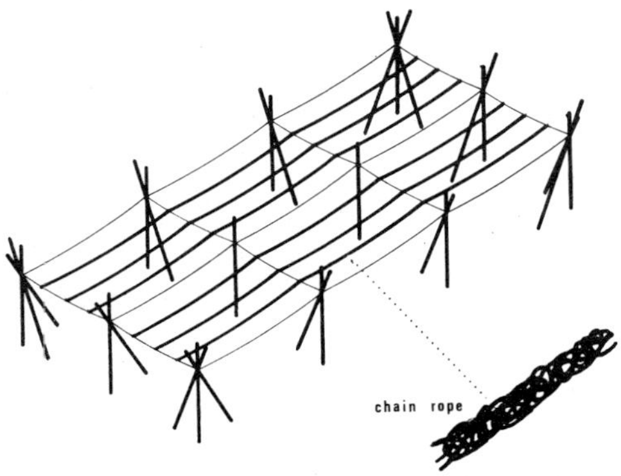

Fig. 77. The vegetable chain-ropes for catching mussel seed are stretched length-wise over a mussel park Italian-style.

and the mussel farmers have to check their parks repeatedly to see whether breakage might occur. For this task, they use a small rake (19 cm across, with four teeth of 15 cm length) to pick up the fallen chain-rope, which has to be attached again to the main ropes of the park.

B.5. Harvesting the mussel seed

From September to November the mussel farmers are busy collecting the mussel seed. They come with their boat, two men on board, and transfer the heavy ropes by hand from the parks to the deck of the boat. The ropes laden with mussel seed are then carried either to the nearby yard at the eastern shore of Vela Draga Bay or to the yard at Pomer, depending on the bay chosen for further growth of the seed. In the latter case, the ropes are transported by van. The clusters of mussel seed are taken from the chain-rope by hand and placed on the filling table before being transferred to cylindrical netting for the next phase of the operation.

B.6. Filling the cylindrical netting with mussels

The Yugoslavian mussels will be grown in hanging culture in the parks of Vela Draga Bay or Pomer Bay exactly as has been described for mussel farming at La Spezia (Italy). For this, they are put in cylindrical nylon netting. Such netting is available in various mesh sizes: 6, 8 and 10 cm stretched mesh, which can be used for mussels of various dimensions. For the seed taken from the collectors one uses the 6 cm mesh the thread of which is 0.5 mm thick. This may seem a rather large mesh considering the size of the mussels, but since they occur in clusters, attached to each other by byssus threads, they will not fall through.

The mussels on the filling table are pushed by hand towards one of the holes in the table leading to a tubular funnel, 30 to 40 cm long, made from plastic or from galvanized iron, over which one slips the cylindrical netting (Fig. 78). The tubes are respectively 6, 8 or 10 cm in diameter, to match the three types of netting. First the end of the netting is closed with a knot tied in the netting material itself, and then the mussels are dropped in until a "sausage" of 170 to 200 cm length has been produced. When seed mussels are used to make the sausage, it takes about 2.5 to 3 kg to fill one length. When the desired length has been filled the top end is closed with a knot. Usually two or three workers carry out this work together and it takes about 5 min to make a 2-m "sausage". The work is done in the yards at Pula or Pomer, or, in bad weather, in the sheds there.

B.7. Hanging the mussel netting in the parks

As soon as is feasible, the mussels, packed in the cylindrical netting (Fig. 79), are brought to the park. A piece of nylon rope serves to tie the sausages

Fig. 78. Filling the cylindrical netting with mussels, making use of a specially-constructed table. (Photo G. Sošić.)

either to the horizontal ropes (coco-fibre rope of 22 mm or nylon rope of 14 mm diameter) as on the Italian-style park, or to the horizontal beams as on the French-style park (Fig. 80). They are attached with 40 cm interspaces. At the outset, the mussels find themselves in an irregular position in the netting, but after a short time in the water they regain their usual orientation, siphons outward. They crawl through the meshes and attach themselves with byssus threads to the shells of neighbouring mussels or to the netting itself, which gradually shifts to a more central position, and is soon out of sight. The mussels grow rapidly, faster in the parks of Vela Draga Bay than in

Fig. 79. Young mussels packed in cylindrical netting to be hung out in a mussel park. (Photo G. Sošić.)

Pomer Bay, presumably because the former are bathed in water enriched by the sewage from the city of Pula.

B.8. Thinning out

The mussels gain in weight, and in due course each "sausage" will weigh 8 to 12 kg (Fig. 81). One should avoid an over-dense population of mussels in the hanging cultures for this may cause clusters of mussels to drop off. Therefore the farmers have to check their parks regularly and take to their headquarters strings of mussels which bear signs of overcrowding or in which they find great numbers of smaller mussels settled in between the larger

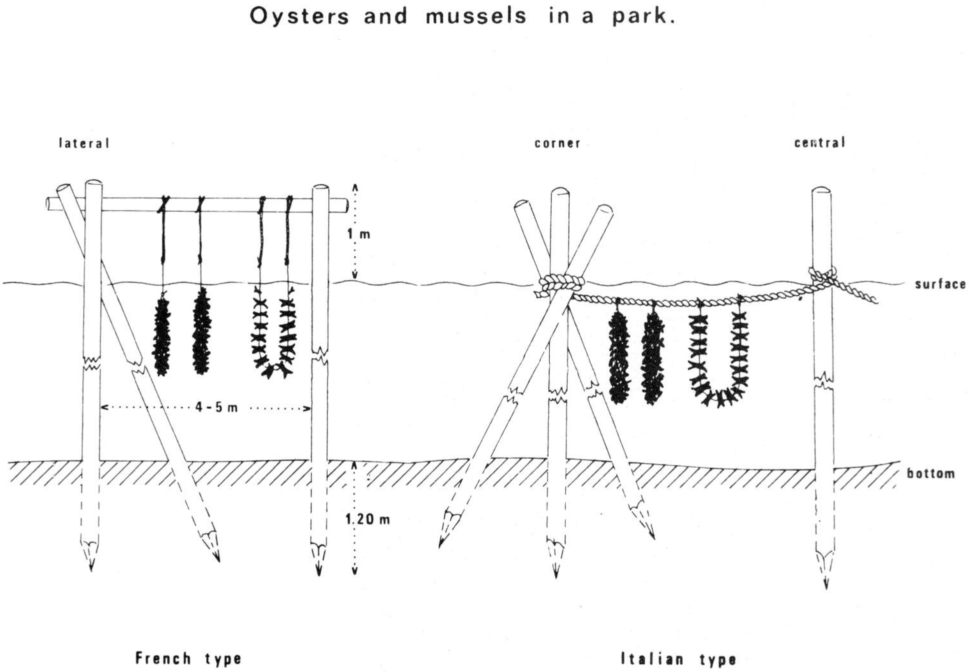

Fig. 80. How the mussel strings are hung out in the French- and Italian-style parks.

ones. The mussels are graded and distributed over a greater length of netting, using now a netting of greater diameter and larger mesh. One should expect to make at least one transfer between filling the netting with seed and harvesting the marketable mussels. Usually the mussels attain the marketable size of about 6 cm shell-length in the course of their second summer, at an age of 18 months and over.

B.9. Cleansing the mussels grown near Pula

Since the harbour of Pula is a source of pollution, because of domestic waste, the mussels grown in nearby Vela Draga Bay may contain infectious microorganisms dangerous to human health. Therefore one never markets mussels directly from there, but always transfers the mussels to parks in Pomer Bay, which is free of pollution. There they are hung from French-style parks, 10 or 11 sausages from each horizontal beam. They are kept there for 15 to 30 days, after which they are bacteriologically impeccable.

B.10. Marketing the mussels

From May onwards mussels have recuperated sufficiently from the strain of spawning activities to be marketed. Sufficient glycogen has accumulated

Fig. 81. The mussels crawl through the meshes of the netting, which can later be found in a central position in the string. (Photo G. Sošić.)

again in their connective tissues to make them palatable. Strings of marketable mussels are now brought from the parks in Pomer Bay to the Pomer headquarters, in response to demand.

Small quantities may be sold on the spot to individual buyers, either by the string or per kg. Any remaining mussels are put in a plastic basket, such as is used for storing oysters, and then into the storage basin. Because of imperfect flushing of the basin they should not be kept there too long, at most a few days.

Larger quantities of mussels are packed in burlap sacks to be taken to the various markets by van, bus or train. In warm weather they will keep their valves tightly closed for about 2 days, in cool weather up to 5 days. In Yugoslavia, the mussel season lasts from May till November; December to April is the spawning season, during which the mussels tend to be lean and impalatable.

C. FARMING RISKS

C.1. Predation

In recent years mussel farming at Pula and at Pomer has met with surprisingly few difficulties. Almost every year the settling of seed on the chain-ropes is heavy enough to stock all the parks, and losses are minimal.

Benthic predators such as starfish and crabs cannot reach the suspended mussels in sufficient numbers to cause concern.

C.2. Competitors

Though suspended oysters and mussels are liable to become smothered by a variety of epibionts, among which ascidians and bryozoa predominate, the mussel farmers of the Pomer enterprise have found little to complain of in recent years. Smothering of the mussels has not occurred and though the French parks would make it easy to hang the strings of mussels in the air for some time to kill shell-less epibionts, this has not been deemed necessary. Mussels carrying many acorn barnacles are difficult to sell, and it is a very laborious job to remove the barnacles from the shells. Therefore late in spring, when they fear a heavy settlement of balanids, the farmers often transfer their mussels from Pula to Pomer if there is enough free space in the parks. For some reason barnacles settle less densely there.

C.3. Parasites

The mussel parasite *Mytilicola intestinalis* does inhabit the coast of Istra. It was indeed in nearby Trieste that this creature was first described. The park of Pula appeared to be free from *Mytilicola*, but at Pomer up to 45% of the mussels were found to be infested, usually with one or two specimens. It can be demonstrated that even such a light infection is the cause of a slight reduction in the quality of the mussels, but since other factors affect growth and fattening, the effect of *Mytilicola* is obscured completely.

The reason why *Mytilicola* does not appreciably harm the mussel in hanging culture is that its larvae have a tendency to seek their host on the sea bed. Samples of wild mussels taken from the sea bed were clearly more heavily infested, which tallies very well with observations made elsewhere in Europe, both in the Mediterranean and on the Atlantic coast.

D. ECONOMIC ASPECTS

D.1. Rents

No rents have to be paid for the sites where parks are constructed.

D.2. Inventory items

The fleet of the Pomer shellfish farm, consisting of five boats, is used both for oyster and for mussel farming. Since about half of the time of the company's workers is devoted to mussel farming one can safely assume that about half the fleet draws upon the mussel farm budget. The fleet consists of five wooden boats, operating both in Vela Draga Bay and in Pomer Bay; three are of the type "Batana Serpi", 6 m long, two of which are equipped with a diesel engine of 7 h.p., and one with a 4 h.p. gasoline engine; the remaining two are flatter boats of the type "Cut-prow Batana" usually towed by the others. The materials used for building the boats are oak, larch-wood and deal. These boats require proper maintenance. For transport on land the company uses a 2-ton van (combi) which is permanently at the disposal of the shellfish farm.

Other inventory items are the wooden shed, measuring 11.0 × 9.60 m, at the Pomer headquarters, a further shed near Pula, and the stone-built storage basin, measuring 13.5 × 6.0 m, at Pomer. There are more than a hundred rectangular plastic storage boxes in use at the headquarters operations. The sheds at Pomer and at Pula should also be reckoned as inventory items which are used about 50/50 for both purposes.

D.3. Expendable items

Since the mussel parks of the Pomer Shellfish Farm cover a greater area than the farm's oyster parks, more poles need to be purchased for their annual repair. Since the mussel parks total about 2 000 vertical and oblique poles with an average life of 5 years, about 400 oak poles have to be bought per year for repair work alone, at a price of Din. 50 each.

The quantity of Sparta grass rope purchased every year is quite impressive: 10 000 kg. This amounts to a total length of 50 to 60 km. It costs Din. 0.40/m, duty and freight included.

The coco-fibre Hercules rope and the 14 mm nylon rope for the Italian-style park last several years, but should be replaced in due course. A certain length of 8 mm nylon rope is required for binding poles together and for supporting mussel strings in the French-style parks.

The cylindrical nylon netting which is used only once is an ever-recurring item of expenditure. It costs Din. 0.30/m, irrespective of the mesh size, and were there no need to thin out the 150 tons of mussels produced each year, about 15 km of netting would be required. Since thinning out is a normal procedure, in reality a far greater length of netting must be bought.

A quantity of fuel has to be purchased for the engines of the boats; for the oyster and mussel operations combined one reckons on about 50 l of gasoline and 500 l of diesel oil per year.

D.4. Personnel

About half the time of the ten workers (five males, five females, plus the three leaders) of the company is devoted to mussel farming. The wages in 1972 were Din. 1 200 for the workers, and Din. 1 500—1 800 for the leaders per month.

D.5. Sales

Mussels sold in small quantities at the headquarters fetch Din. 3/kg, but the wholesale price is Din. 2.50/kg. The freight has to be paid by the purchaser. On the home market, the mussels fetch Din. 3.50 to 6/kg.

The total production of mussels from the Pomer shellfish farm is reported to amount to 150 tons. It would be possible to produce more, but at present the local markets cannot absorb more.

E. GOVERNMENTAL ASSISTANCE

Since the Pomer Shellfish Farm is part of the Agricultural Cooperative "Poljoprivredna zadruga '1 maj' ", the Government exercises a high degree of control over the farm and determines the scope of its operations now and in the future. No rent is paid to the owner of the grounds (the State) but if the budget shows a deficit, the economic adviser of the large State Cooperative decides what steps should be taken. From outside the Cooperative the shellfish farmer can rely on the assistance of the biologists of the "Ruder Bošković" Institute for Marine Research at Rovinj. Staff members of this Institute have carried out an appreciable amount of research on the hydrographical and biological conditions in the various shellfish areas of the Istra peninsula.

The "Ruder Bošković" Institute for Marine Research at Rovinj has already carried out an appreciable amount of work on the biology of the mussel and on its parasites and competitors. Such investigations can form a basis for methods of control when serious problems arise. One can also rely on the continued interest of the Institute in all hydrographic and biological problems related to the shellfish industry in Istra.

F. GENERAL IMPRESSION

Mussel farming Yugoslavian-style has much in common with the system of rack culture practised elsewhere on the Mediterranean coasts. New materials such as nylon rope and netting have been adopted and it is to be expected that more durable material will in due course be applied generally in the construction of mussel parks.

Hydrographic and biological conditions are more conducive here to mussel farming than in many other sections of the Mediterranean coast. The multitude of islands and inlets on the Yugoslavian coast offer the degree of shelter required for this method of farming, and water temperatures and plankton content are such on the coast of Istra, that a more rapid growth and fattening of mussels occur than on almost any other part of the Mediterranean coast. Since many of the waters are completely unpolluted, a product is delivered which is safe to eat raw. If growth is more rapid in water enriched with some sewage, the marketable mussels can be transferred to clean water, in which a short sojourn (about 2 weeks) will suffice to make them bacteriologically acceptable. The production of seed is almost fully under control since the chain-ropes made from Sparta grass yield excellent results. That the total production of marketable mussels is still rather modest is a result of modest demand. Since several factors, among which pollution predominates, are leading to a decline of mussel production in neighbouring Italy, it might be possible to find a good market there in due course. At the present price, it will be easy to compete with mussels imported from Spain.

G. BIBLIOGRAPHY

Hrs Brenko, M., 1965. Contribution à la connaissance du copépode parasite *Mytilicola intestinalis* Steuer, particulièrement dans l'Adriatique. Rapp. P. V. Réun. Cons. Perm. Int. Explor. Mer, 18: 168—170.

Hrs Brenko, M., 1967. *Mytilicola intestinalis* Steuer (Copepoda parasitica), a parasite in mussels in the East Adriatic. Thalassia Jugosl., 3: 143—155.

Hrs Brenko, M., 1967. Index of condition in cultured mussels on the Adriatic coast. Thalassia Jugosl., 3: 173—178.

Hrs Brenko, M., 1968. Biometrical analysis of the mussel *(Mytilus galloprovincialis)* along the eastern coast of the Adriatic. Thalassia Jugosl., 4: 19—30.

Hrs Brenko, M., 1972. The Relationship between Reproductive Cycle and Index of Condition in Mussel (*Mytilus galloprovincialis* Lmk.) in the North Adriatic Sea. Rep. Gen. Fish. Counc. Mediterr., Colloque Athens, 5 pp.

Hrs Brenko, M. and Božić, E., 1965. Influence of the parasite *Mytilicola instestinalis* Steuer on mussels. Gen. Fish. Counc. Mediterr. Tech. Pap., 18: 181—183.

Hrs Brenko, M. and Calabrese, A., 1969. The combined effects of salinity and temperature on larvae of the mussel *Mytilus edulis* L. Mar. Biol., 4: 224—226.

Hrs Brenko, M. and Igić, L., 1968. Effects of fresh water and saturated sea-water brine on the survival of mussels, oysters and some epibionts on them. Stud. Rev. Gen. Fish. Counc. Mediterr., 37: 29—44.

Marinković-Roje, M., 1968. Sezonske varijacije sastava dagnji *Mytilus galloprovincialis* Lmk. iz Limskog kanale. Thalassia Jugosl., 4: 69—83.

Chapter 7

MUSSEL FARMING RUSSIAN-STYLE IN THE BLACK SEA

Example
Experimental Mussel Farming by AzCherNIRO, 334502 Kerch, U.S.S.R.
Period of observation: 1969—1973.

A. BACKGROUND

A.1. General principles

The shallower parts of the western and northern sections of the Black Sea are remarkably rich in natural mussel beds. Beyond a zone of sand and rocks one may find such natural mussel beds on a rather muddy sea bed down to a depth of 65 m. It is especially the northwestern section, the area off Odessa and Ochakov, which accommodates rich and extensive natural mussel beds, but one also finds such beds on the Crimean coast, off Kerch Peninsula and south of the Danube delta along the coasts of Rumania and Bulgaria.

Since these natural beds are a potential resource of valuable animal protein, they have been opened to commercial exploitation. It is estimated that the entire mussel stock in the Black Sea amounts to about 10^7 metric tons, of which about one-third is of commercial size. Not all these beds are readily accessible, but since at least one-tenth is within easy reach of the Russian fleet it must be possible to harvest a sustainable annual catch of the order of 100 000 metric tons.

The commercial exploitation of the natural mussel beds of the Black Sea is concentrated in the area Odessa—Ochakov with headquarters at the State Fishery and Processing Enterprise at Ochakov. Harvesting is done by dredging (Fig. 82) and the season lasts from October 1st till the end of May. The official regulations for harvesting marine plants and aquatic invertebrates in the territorial waters of the U.S.S.R. prescribe a minimum legal size of 50 mm shell-length for mussels, and the State Fish Resources Protection Agency, supervising the harvesting on the natural beds, sees to it that no undersized mussels are landed. Since mussels of all sizes may occur on the natural beds the catch must be sorted out on board and all the undersized mussels returned undamaged to the beds. Therefore a production line is operated on board the ships. Nowadays there are two ships of the S.Ch.S.-150 class (medium size Black Sea seine-net ships) which provide the processing plant at Ochakov with mussels, operating on a scheme ensuring a regular supply. Mussels form only one of the raw materials worked by this plant. The ships of

Fig. 82. Dredging mussels on the natural beds in the Black Sea. (Photo A.I. Ivanov.)

the S.Ch.S.-150 type are equipped with a 150 h.p. engine and have an auxiliary diesel generator of 25 kW to operate the fishing and sorting devices. The mussel dredge used consists of a rectangular steel frame, equipped with a sharp knife on both of the long sides, these knives making an angle of 60° with each other. Such a double-sided dredge offers the advantage that it will also operate when it lands upside down on the mussel beds. A bag of wire netting with 30 mm meshes is attached to this frame. The dredge is lowered and hoisted with the aid of a powered winch. When a dredge load of mussels has been swung inboard the dredge is not dumped in the way the Dutch mussel fishers operate, but the sides of the bag are opened with a special device when a lever touches the sidewall of the hold. Since lowering, hoisting and emptying of the dredge require quite an appreciable amount of time and labour, the Central Design and Technical Bureau AzCherRyba (Sea of Azov and Black Sea Main Fisheries Administration) has developed a method of ensuring continuous harvesting, which operates even in rough weather and is considerably less dangerous for the workers than repeated hoisting of the dredge. In this method a long suction hose is used, leading from a RBU-150 fish pump on deck to the interior of the bag of the mussel dredge. To fish on beds of 6 to 20 m deep a length of dredge-line and suction hose of 24 to 46 m is needed. The boat operates at a speed of 3 knots, or when the beds are densely populated, about half that speed. Mussels and water come aboard in a ratio 1 : 10 and after passing a water separator and a shock-absorbing trough the mussels are guided to the production line. The on-board production line is in principle comparable with the installation the Dutch mussel dealers use ashore, the main parts being a washer and a blower. The washer consists of a rotating cylinder constructed as a series of rings made from round iron bars, the interspaces arranged in such a way that the undersized mussels will fall through together with silt and sand washed off by a powerful jet of water. The washer measures 100 cm in length and 50 cm diameter. It is installed horizontally but a spiral-shaped internal ridge with a 24 cm pitch ensures forward movement of the mussels in the rotating cylinder. A rubber strip installed above the cylindrical rotating washer pushes back any mussels which become lodged between the iron rings. Undersized mussels and all the adhering sand and mud are thus eliminated and washed overboard down a sloping chute, but many empty shells may still be found among the mussels. They are therefore taken by a conveyor belt to the blower. This is merely a metal grid under which a powerful air pump has been installed, delivering 9 000 m^3/h air. The empty shells are thus mechanically eliminated and the clean and sorted mussels, ready for delivery at the processing plant, are collected at the end of the production line.

All the mussels harvested this way, with or without the fish pump, are brought to the processing plant where a variety of canned products for the home markets are made. As such are listed: "natural mussels", "Black Sea mussels in their own juice", "plov de moules à la Taurida" ("plov" is a rice—mutton dish very popular in Central Asia), "mussels with buckwheat porridge", "Solyanka of mussels and fish" ("Solyanka" is a dish with pickled vegetables) and "smoked mussels in sunflower oil".

If one considers the system of mechanized harvesting and processing of mussels from the natural beds in the Black Sea, knowing also that these beds are still largely underexploited, it is surprising to learn that the AzCherNIRO (the Azov-Black Sea Research Institute for Fisheries and Oceanography) and, more specifically its section for research on commercial invertebrates guided by the biologist A.I. Ivanov, does its utmost to develop a system of mussel farming adapted to the local hydrographic conditions.

As a general rule one can state that where a natural resource is exploited by fishing, and a good supply is landed, farming, which inevitably means investment of a considerable amount of money and labour, cannot be remunerative since the farmed product has to compete on the market with the wild stock caught by the fisherman. Still, there are very good reasons for the efforts of AzCherNIRO and the results obtained thus far indicate that a system of mussel farming adapted to the local conditions may become an economically sound enterprise.

The fishing of mussels on the natural beds in the Black Sea is quite a costly method, especially as all the mussels have to pass a production line on board. The dredge used is rather small (the knives measuring less than 1 m in length) and instead of the 4 dredges with knives of 1.60 m long with which a Dutch mussel ship is equipped, one dredge only is operated. The natural mussel beds may be well stocked with up to 7 or 8 kg/m^2 of mussels, but unlike the Dutch-style mussel park, only a fraction of the mussels are of marketable size and elaborate washing and sorting on board must be carried out. The harvesting and processing operations on board require six workers, but the net yield is not more than about 1 000 kg/h of marketable mussels. A Dutch mussel ship operating with a crew of three, fishes more than 10 000 kg/h from a farming plot, and its hold can contain over 50 000 kg of mussels. The production line on board the Russian vessels can handle about 1 000 kg/h of mussels, whereas a similar type of production line ashore processes about 5 000 kg/h in The Netherlands. The limited efficiency of fishing and the large number of workers required on board, raises the production costs for the wild mussels landed at Ochakov to as much as R 60 to 90 per 1 000 kg.

That the total quantity of marketable mussels annually landed and processed in the State Fish Processing Plant at Ochakov amounts to 1 500 to 2 000 metric tons, rarely more, although the natural beds would allow an annual catch of some 100 000 metric tons, is not, however, to be ascribed to the small number of boats operating, to the limited quantities harvested per fishing day or to restricted marketing possibilities, but rather to the occurrence of pearls in the mussels. The mussels from the natural beds in the Black Sea often contain pearl-like calcareous concretions which render them unacceptable to the consumer. All processing operations in the cannery could be mechanized with the exception of one step: eliminating mussel meat containing pearls, a task which is still done by hand. This is the bottleneck in the processing scheme which keeps the total quantity of mussels produced small. The occurrence of pearls is in all probability due to the impact of a Gymnophallid trematode. The same phenomenon has been observed in mussels from natural beds on the Atlantic Coast of the United States and was the cause of failure in marketing mussels as human food during World War II. Some kind of waterfowl is probably host to the adult phase of the trematode, but before considering a possible method of control one should study in detail the reasons for pearls being abnormally abundant in Black Sea mussels.

It is above all the absence of pearls in mussels grown in hanging culture which makes it so attractive to farm mussels in the Black Sea. If pearl-free mussels were available, the processing plant could in all probability handle about ten times the quantity of mussels processed today.

An additional advantage of the mussels farmed in the experimental plant is their higher meat content. The mussels from the natural beds yield after steam treatment only about 8% in meat weight, whereas the farmed mussels yield 11 to 14%, a noteworthy difference, which also makes farming attractive. When all these factors were combined, viz. the high cost of harvesting mussels from the natural beds, the better growth and fattening of cultivated mussels, and above all the absence of pearls in mussels produced out of reach of the sea bed, the economists of the Fisheries Institute were led to the preliminary conclusion that farming would not only mean production of mussels of better quality, but would also produce mussels at a lower price than those at present harvested from the natural beds, perhaps even half the price.

This explains the paradoxical situation in which one initiates a farming procedure in the presence of rich natural beds, and may in due course achieve considerable economic success. But before farming takes over completely from fishing some practical problems have to be solved.

A.2. Biology of the species

The general biology of the mussel is referred to in the chapters on mussel farming in The Netherlands, Spain and Italy. The mussels of the Black Sea are usually reckoned to belong to the species *Mytilus galloprovincialis* and more specifically to its variety *frequens*, but it differs from *Mytilus edulis* L. so little that no special treatise on the biology is needed, and doubt may be expressed whether one should really speak of two different species.

The mussels of the Black Sea have generally rather thin shells, especially where the salinity is low and where currents and wave action are of little importance. The mussels of the Black Sea can occasionally reach a shell-length of 11 cm, but the bulk of the adult mussels on the natural beds measure 5 to 6.5 cm in shell-length and weigh from 13 to 14 g each. There are indications that the mussels of the natural beds in the northwest of the Black Sea require about 30 months to reach the minimum legal size of 5 cm shell-length, but in the farming experiments in the Strait of Kerch it was observed that 5 cm shell-length may be reached within 18 months.

In order to study the occurrence of mussel larvae in several stations, hundreds of plankton samples have been taken by making a vertical haul with a Jeddy net of 25 cm diameter. The data obtained show that reproduction begins quite late in this area, which is not surprising taking into account the prevailing water temperatures here. Though a few larvae are found as early as April, the bulk makes its appearance in June and July, and only few were seen in the samples taken later than September. Little is known about the exact duration of the planktonic phase, but experimenting with collectors made it clear that one should not wait till the number of pelagic larvae is high to put the collectors to sea. Collectors immersed long before that time appeared to catch a better crop of mussel seed than those offered in summer. This tallies very well with the experience of the mussel farmers in France, i.e., that smooth and clean collectors only become attractive as a settling surface to mussel larvae after having been covered with hydroids or other filamentous organisms.

The natural beds of mussels are predominantly found on sandy mud, but where rocks or wrecks occur those may also be found to carry a heavy load of mussels, which cannot, however, be harvested economically. The density of mussels on the natural beds varies from place to place and from year to year. Locally one finds up to 890 individuals per m^2 with a biomass of 7 278 g. Considering that the shells of the Black Sea mussels are thinner than those on the Atlantic coast of Europe, these figures come very close to those found for well-stocked mussel parks: 500 adult mussels or 1 000 half-grown mussels per m^2 with a biomass of 8 to 10 kg.

A.3. Geographical situation

The AzCherNIRO headquarters are to be found in the city of Kerch on the eastern-most tip of Kerch Peninsula, which is continuous with the Crimea. Its exact geographical position is 45° 22′ N, 36° 27′ E.

Experimental mussel farming is predominantly carried out in the Bay of Kerch, 1.6 miles in length and 1.2 miles across, situated southeast of the city. The exact place where the mussel parks were constructed is indicated on the map (Chart 9). Further farming experiments were carried out in various places in the Strait of Kerch, which connects the Sea of Azov with the Black Sea (the exact sites indicated on the map) and also in the vicinity of Ochakov (46° 37′ N, 31° 33′ E) and in some sheltered places in Yagorlytskiy Bay and Tendrovskiy Bay, south-southeast of Ochakov, all indicated on the map.

A.4. Hydrographic pattern

The Black Sea has an unusual hydrographic pattern. Its communication with the Mediterranean Sea and thus with the Atlantic Ocean is narrow, and the volume of high salinity water which creeps into the Black Sea along the bottom of the Dardanelles is so small that there is very little flushing of the deeper layers of the Black Sea. Since considerable quantities of organic matter sink down to these deeper layers where the sunlight cannot penetrate, the oxygen supply has been exhausted for aeons, and at depths over 180 m poisonous hydrogen sulphide accumulates, making the existence of all higher forms of life impossible. This, however, is no problem for the mussel indus-

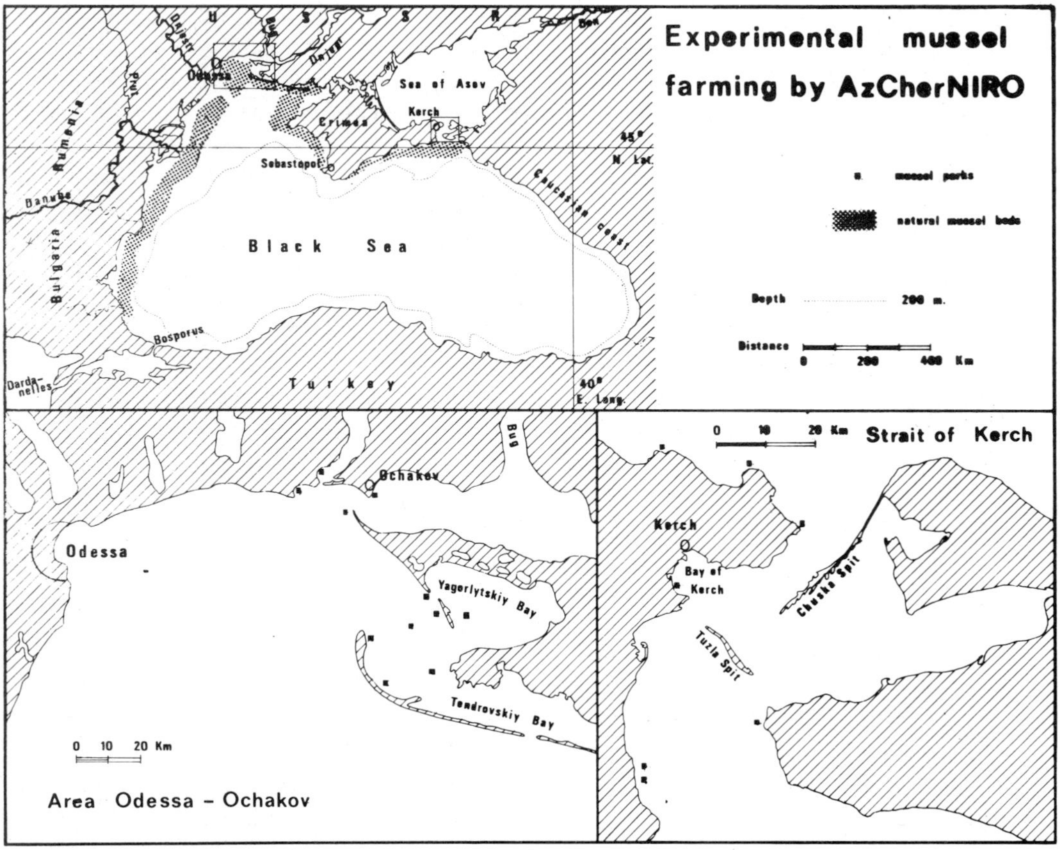

try since the natural mussel beds occur in the shallow coastal area, and the experimental farms are set up in still shallower sites.

Tides, currents and waves. The narrow passages to the open ocean lead to almost negligible tides in the Black Sea. Even so, the superficial water layers are not stagnant here. The prevailing winds lead to a certain circulation pattern and at the southern shore of the Crimea one can expect in the summer season an east—west current with a velocity of 5 to 10 cm/s. In the Strait of Kerch, where the most important experiments are carried out, the direction in which currents will run is largely determined by the wind. The winds often blow quite strongly in this area, resulting in swifter currents and higher waves, which may damage mussel parks constructed in poorly-sheltered sites. At Kerch, the number of days with wind force 5 and over amount to 18—19 in winter, 15—19 in spring, 11—13 in summer and 14—17 in autumn.

Depth. The Strait of Kerch connects the Sea of Azov with the Black Sea, and separates the Crimean Peninsula from the Caucasian coast. The Strait is about 22 miles long from north to south and in its narrowest part only 2.5

miles wide. At its northern end the Strait is up to 10 m deep, in its southern section up to 18 m, the central part having a depth of 5.5 m or less. Kerch Bay, in which the first experimental mussel park was constructed, is about 4 m deep with a bed of muddy sand. The currents govern the nature of the bed, which is sandy in those places where stronger currents prevail. Remarkable features of the Strait of Kerch are the two sand spits, the Tuzla spit across the Strait in its central part, and the long Chuska spit in the northeastern section. The sea bed is sandy in the vicinity of these.

Salinity. In the deeper layers of the Black Sea (over 300 m) the salinity mostly shows a constant value of about 22‰, much lower than the Mediterranean waters, due to the continuous influx of river water and the relatively small amounts of high salinity water creeping in through the Dardanelles. In the upper water layers the salinity of the Black Sea is lower still, varying between 17 and 18‰ in the layers above the thermocline which is found at a depth of about 50 m in the central part, and at 70 to 80 m closer inshore.

It is especially the rivers Danube, Prut, Dnjestr, Bug and Dnepr which discharge so much fresh water into the Black Sea, that it governs the salinity in the superficial water layers. The river Don flows into the Sea of Azov and may discharge so much fresh water that the salinities may drop to 9‰ in the Strait of Kerch when the currents come from the north. Since the outflow from the Sea of Azov prevails over the influx from the Black Sea, especially in spring and early summer, one often observes a north—south current in the Strait of Kerch, accompanied by lower salinities.

Water temperatures. Though the latitude is about 45° in the area concerned, water temperatures may drop to low levels in winter and sometimes the Strait of Kerch may become covered with ice. The Sea of Azov becomes frozen more often and in spring large masses of ice will find their way through the Strait of Kerch to the Black Sea. The poles of the mussel farms, if not taken away in time, would be destroyed by the floating ice masses.

The water temperature rises gradually in spring, but is often only 4°C early in April. In the first half of May one may observe a temperature of 12° C, and in mid-June 21°C. The temperature may reach 24°C in mid-summer.

Oxygen level. Though the well-known hydrogen sulphide-containing water masses of the Black Sea are too far below the surface to affect either the natural mussel beds or the experimental mussel farms, the oxygen content is not always optimal. When the Straits and Bays are covered with ice in severe winters, oxygen may become depleted under the ice cover, especially when a layer of snow on top of the ice intercepts the sunlight, preventing carbon dioxide assimilation by the phytoplankton. Another danger is prolonged calm weather in summer which locally may also result in low oxygen values of the water.

Phytoplankton. Many observations have been made on the phytoplankton since this constitutes the food of the mussels. That the pattern of the wind-dominated currents is rather irregular here, is a disadvantage for the mussels, for where a dense population of mussels occurs, either on a natural bed or on a farm, the mussels should be supplied continuously with water rich in plankton. Therefore areas with strong tidal currents are more appropriate for mussel farming than waters which are often almost stagnant, even if one were to find that the quantity of phytoplankton per m^3 of water was about the same.

In the Sea of Azov, which receives so much material from the river Don, there is often quite rich phytoplankton, combined with detritus. When such water is carried to the mussel parks good feeding may result, despite the lower salinities.

Observations made by the staff of the AzCherNIRO Institute in 1970 showed values of 100—500 mg diatoms per m^3 in spring, with a peak figure of 7 400 mg/m^3 late in March, values of 20—70 mg/m^3 in summer and again somewhat higher figures (50—400 mg/m^3) in autumn. For dinoflagellates, the figures recorded were 1—40 mg/m^3 in spring, 50—300 mg/m^3 in summer, with a noteworthy peak of 1 630 mg/m^3 late in June and 80—130 mg/m^3 in autumn. Species of *Skeletonema*, *Chaetoceras*, *Thalassionema*, *Coscinodiscus* and *Cerataulina* predominate among the diatoms, and species of the genera *Exuviella*, *Prorocentrum*, *Peridinium* and *Glenodinium* among the dinoflagellates.

A.5. Legal aspects

Various legal aspects must be taken into consideration to get a good understanding of the mussel farming venture in the Black Sea. All the waters with their natural resources are the property of the State. If an agency of the State such as AzCherNIRO constructs experimental mussel farms in sites deemed to be favourable, it need not pay any rent or fee to the State. The fishing on the natural beds is also organized and supervised by the State. Therefore one need not fear a clash between fishing on the natural beds and the mussel farming enterprises. The economists of AzCherNIRO will study the qualities and quantities produced under both systems and will duly take into consideration capital investments and labour costs and will ultimately advise on the best path to follow in the future. Though the natural resources belong to the State and though State enterprises have the monopoly in exploiting these resources, the citizens of the U.S.S.R. may collect some mussels from the natural beds free of charge. A bag-limit of 8 kg of mussels per person per day has been set by law.

The regulations for harvesting marine plants and aquatic invertebrates in the waters of the U.S.S.R. stipulate the minimum legal sizes. For the Black Sea mussels the minimum legal size is 5 cm shell-length. It is the State Fish Resource Protection Agency which supervises the fishery and sees that no undersized mussels are landed. To avoid exploitation when the mussels are lean as a consequence of the spawning season, the official mussel fishery season runs from October 1st till the end of May.

The State Sanitary Inspection also supervises the hygiene aspects of aquatic products. Though most of the mussels consumed in the area under consideration undergo some sort of cooking procedure, they are occasionally preferred raw. Therefore good bacteriological control is essential where mussels are harvested from sites at which pollution by domestic sewage may occur to some extent.

B. SEQUENCE OF OPERATIONS

B.1. Preliminary investigations

If mussel farming is proposed in an area where no practical experience exists, one should first of all investigate where and when the settling of mussel seed is to be expected. If there are no natural resources from which mussel seed can be taken as basic material for a farming enterprise, such investigations are a prerequisite, for haphazard placing of artificial collectors would be too costly and too time-consuming as a means of detecting the best sites and the best time of the year to collect mussel seed. The AzCherNIRO scientists decided, therefore, to collect plankton samples from early spring till the onset of winter in a large number of inshore stations, especially in sheltered bays where a farm would be safe from gales. A 12 m motor boat equipped with a 150 h.p. engine and operating with a crew of three is at their disposal. From earlier investigations it was known how to identify mussel larvae in Black Sea plankton samples and the AzCherNIRO biologists are satisfied that the results obtained give a clear picture of the whereabouts of the mussel larvae and of the periodicity in their appearance. Small numbers (less than 10 larvae per m^3) were observed in early spring, the first ones making their appearance shortly after the water temperature exceeds the 4°C level which is considered as being critical for reproduction in *Mytilus*, but in the course of the month of May the number of larvae rose steadily to reach peak values up to 10 000 larvae per m^3 in June and July. In Kerch Bay the number of larvae in the plankton dropped steeply thereafter, but in stations in the northwestern part of the Black Sea, August and September also revealed reasonable numbers of larvae.

Though these investigations do reveal where settling of mussels could be expected, or rather where it will in all probability not take place on a commercial scale, they cannot be used for selecting the best time for offering collectors to the settling larvae. This is not because of ignorance about the duration of the planktonic life of mussel larvae, for one could measure the shell-length of the larvae observed in the plankton samples and one knows at which size they settle. The reason is that unlike oyster larvae, which require clean collectors, recently immersed, mussel larvae do not readily settle on the clean smooth surfaces of artificial collectors but prefer a substratum provided with filamentous objects such as hydroids or some red algae. This was fully confirmed by the experiments of the AzCherNIRO biologists which led to the most profuse settling in June, on collectors immersed in the parks as early as April.

B.2. Constructing an experimental mussel park

The first experimental mussel park was set up in the Bay of Kerch in the year 1969. This site was maintained for the experiments in the following years, the parks having to be removed in winter and to be reconstructed after the disappearance of the ice. In the year 1973 six more parks were built in the Strait of Kerch.

The park in the Bay of Kerch (Fig. 83) covers 500 m^2 and its framework consists of two rows of six poles, the poles being placed vertically 10 m apart, with 10 m between the two rows, hence an area of 500 m^2. In the site used in the Bay of Kerch, the depth is 4 m; near Ochakov, where another experimental park was set up, the depth is 6 m. The poles are of pine, 10 to 15 cm in diameter and of such a length that their tops stand about 2 m above

Fig. 83. The experimental mussel park of AzCherNIRO in the Bay of Kerch. (Photo A.I. Ivanov.)

the water surface. They are kept in position with the aid of 6 mm steel wires and 16 anchors.

Construction of the framework of a park takes 1 or 2 days, and is carried out by a team of fishermen, members of the local Cooperative Fisheries Enterprise, as agreed by the management. The poles are placed on, and not driven into the sea bed. The lower end of each pole is pointed so that they gradually sink into the sea bed under the weight of the mussels. Big stones attached to the lower ends of the poles serve temporarily to counteract their buoyancy, but when all the anchors are in place and the top ends have been interconnected with steel wires from pole to pole and from row to row, SCUBA divers detach the stones. For a 500 m^2 park of 12 poles about 350 kg of 6 mm steel wire and 16 anchors are needed. The poles need not be protected against attack by shipworms (Teredo) with tar or creosote since they are required for one season only, the parks being removed before ice can destroy them. Concrete poles are, however, under consideration for future experiments.

Once the parks are in position 10—12 mm kapron cords are stretched from pole to pole and from row to row just above the water level. These cords will serve as supports for the collectors. Due to sagging, the middle sections will become submerged in due course.

In recent years steel supports have also been used in some sites for hanging out experimental collectors. These supports, originally developed for the hibernation of mussels produced in the parks, can be used throughout the year. Their construction will be described below in the section dealing with hibernation.

B.3. Making the collectors

Various materials have been tested as collectors for mussel seed. The workshop of AzCherNIRO makes plates of standard dimensions (8 × 3 cm) and inserts these, 8 cm apart, in a kapron cord. Then two 3 mm kapron lines are twisted around one another, starting with two lengths of 50 m and the plates are inserted while twisting. When ten plates have been inserted (Fig. 84) the twisting is continued to produce a length of rope from 1.5 to 2 m without the plates, and then a new series of ten plates is inserted. Thereafter the collectors are cut to the desired length, leaving free ends for tying to the supporting ropes in the parks and for attaching a sinker (a piece of iron of about 0.5 kg) to the lower end.

Cement and foam plastic have been tried as material for the plates. In the former case, plates of 8 × 3 cm were sawn from 0.8 cm thick commercial roofing slates, and in the latter case plates of the same standard dimensions were cut from blocks of foam plastic (PVC). The latter material appeared to be a better substratum for mussel seed than cement, presumably because its rougher surface encourages the settling of hydroids. Some 9 000 plates can

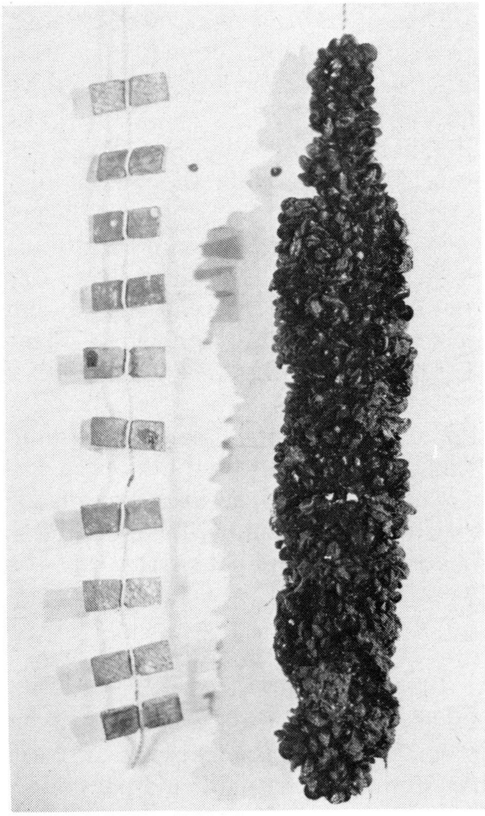

Fig. 84. String of collectors made in the workshop of AzCherNIRO, and a similar string after having been hung in the experimental mussel park till the month of September. (Photo A.I. Ivanov.)

be cut out of 100 kg of commercial foam plastic, which costs R 1.24/kg. One worker equipped with an electric saw can produce about 2 000 plates per day and a team of four workers can make as many as 120 collectors of ten plates each per day. The AzCherNIRO annually uses some 4 000 collectors for its experimental mussel farming.

B.4. Bringing the collectors to the park

The collectors, prepared in the workshop, are transported to the park on the 12 m research ship. The personnel of the research institute itself bring out the collectors and attach the collectors to the horizontal kapron lines of the park. The longer free end of the collector is tied to the horizontal line, with interspaces of 25 to 30 cm between collectors. It takes only 14 s to attach one collector to its supporting line. The collector, weighted by the sinker, hangs down vertically and the topmost plate of each collector is intentionally kept 2 m below the water surface, for experience has already shown that settling is poorer in the topmost water layers.

Since mussel farming is still in the experimental phase here, not all the collectors are put out in the period tentatively considered to be the most appropriate, in an effort to get as rich a settling as possible, but rather, batches of collectors are placed at 10-day intervals, starting in February and continuing till December. The number of collectors in each batch is so chosen that one can remove two collectors from every batch at 10-day intervals till the end of the current year, and thereafter three times (March, September, October) during the next year (in the Strait of Kerch) or six times during the next 2 years (in the northwestern Black Sea).

Thus, although the maximum quantity of marketable mussels will certainly not be produced, knowledge of, and insight into, the fluctuations in settling and growth in various experimental sites is acquired, and this knowledge is gained during years with somewhat different environmental conditions. Such information can produce a sound basis for setting up mussel farms of a truly commercial character.

B.5. Assessing the results

At 10-day intervals, collectors are taken up from the park for analysis ashore (Fig. 85). It takes about 25 s to detach a collector from the horizontal supporting ropes. Ashore, the number of mussels per plate and per collector are assessed and extrapolated to numbers per m^2. The growth pattern of the mussels is studied, using shell-length as a practical parameter, and the meat content determined by simply cutting the meat from the shell and/or by weighing the meat collected after steam treatment. Thus an impressive amount of data has been collected showing that the best settling occurs at a depth of 2 to 3 m; foam plastic gives a higher number of mussels per m^2 than cement plates; collectors put out in the Kerch Bay park in April appeared to carry considerably larger mussels than collectors put out later in the year; where collectors were put out in spring, a shell-length of 3 cm can be expected in October; a shell-length over 5 cm, the minimum legal size, can be reached after about 18 months growth. The latter figure holds good for Kerch Bay with rich phytoplankton coming from the Sea of Azov, whereas it was concluded from the data obtained in the experimental parks in the northwestern Black Sea that the mussels there might require about 30 months to reach the minimum legal size.

The number of mussels per m^2 varied enormously from place to place and according to the time of year that the collectors were put out. In Kerch Bay numbers as high as 34 000 mussels per m^2, representing 73 kg of biomass per m^2 were found. Expressed per collector, this means a harvest of 20 to 30 kg of

Fig. 85. In due course the strings hung out in the park are well covered with mussels which have crawled out over the whole length of the ropes. (Photo A.I. Ivanov.)

mussels, which in due course crawled out over the whole length of the collector.

Not only are the experiments designed to provide this type of information on the best time for putting out collectors, the most appropriate depths for settling, and the growth and fattening in various places in different years, but collectors well-laden with mussels are also used to experiment with the mechanical harvesting of mussels from the collectors, experiments which have already led to positive results. The State Processing Plant at Ochakov also studies the qualities of these mussels for the canning industry. It has already been demonstrated that the cultivated mussels are free of pearls and have a better meat content than those harvested from the natural beds.

The economists of AzCherNIRO are of course very interested in the results and try to gain estimates of the production costs per kg of farmed mussels. One should, however, be careful in extrapolating from a few best figures, and also in extrapolating the results from a 500 m^2 farm to hectares, for food has to be brought in by the currents, and a hectare densely hung with mussel collectors may show a very different end-result compared with a 500 m^2 park with the same population density.

The expectation based on extrapolation, is that 1 ha of park might contain at the end of the first year about 600 metric tons of young mussels with a meat content of about 74 tons, increasing in the second year to 4 000 metric tons with a meat content of 500 tons, but this still seems rather optimistic.

B.6. Hibernation experiments

Ice and gales threaten to destroy the parks in winter, and since only the fastest growers of the earliest settlers may have reached the minimum legal size at the onset of winter, it is necessary to develop a good system for hiber-

Fig. 86. Metal frames equipped with floats and sinks are also used for hanging out collector strings. The same frames are used for hibernation experiments in deeper water. (Photo A.I. Ivanov.)

nation so that the mussels can make use of a second growth season to reach marketable size. In the northwestern section of the Black Sea, which is poorer in phytoplankton, two more growth seasons may be required. Since the parks should be taken away towards the end of the year metal frames are constructed, each about 4.5 m in length, from which 24 collectors can be hung, loaded with half-grown mussels (Fig. 86). The collectors can be kept in a vertical position at sufficient depth to keep them free from ice and waves, by attaching foam plastic floats (PVC) to the topside of the supporting metal frames, and by attaching a sinker, (a hemispherical pig of cast iron of about 300 kg) to the collectors' lower ends. Thus the frames carrying the collectors stand vertically in mid-water layers in a well-selected site. SCUBA divers can pick them up again when the dangers of winter are no longer a threat. The collectors can then be brought to newly-constructed parks or be kept on the steel frames if so wished.

By these methods the growth of the farmed mussels has been studied into their second and third years.

B.7. Further work

Efforts will be made to develop a scheme which profits as much as possible from the early part of the settling season so that the mussels will have attained a fair size by winter, when growth diminishes. Still, there is little hope that the mussels will by then have reached commercial size, not even in the Strait of Kerch where the water is richer in phytoplankton than elsewhere in the Black Sea mussel district. This makes it clear that some hibernation technique may be needed to carry the crop over to the next growing season and it is still questionable whether the metal frames now in use on an experimental scale are technically and economically acceptable, the more so since heavy winter losses have already occurred with these devices.

Another possibility would be to process small mussels under one year old, though this would require a change in the regulations which now stipulate 5 cm minimum shell-length.

Considering the modest meat content of the Black Sea mussels compared with those from the Atlantic coast of Europe, research on feeding behaviour in the mussel is being considered, with the hope of increasing the quantity of mussel food. It should, however, be remembered that all the molluscan shellfish produced commercially in the world still live entirely on a natural diet. It will certainly be no easy matter to develop an artificial diet for these invertebrates on an economic basis.

C. FARMING RISKS

C.1. Hydrographic conditions

The greatest menace to mussel farming on the northern coast of the Black Sea is ice. Not every winter will there be a complete sheet of ice covering the sites where the experimental farms are set up in the Strait of Kerch, but large masses of ice may drift through the Strait in spring, coming from the Sea of Azov and out of the river Don and its tributaries. No parks constructed of poles and wire, nor rafts, could withstand the combined impact of ice, currents and wind. Without a good system of hibernation the mussel venture cannot lead to consistant remunerative results. In recent winters considerable quantities of mussels have been lost from the collectors hung from the steel hibernation frames. It is evidently not only the direct action of ice and waves which does the damage, but when the environmental conditions become unfavourable for the mussels (e.g., the oxygen content of the water becomes too low because of the ice coating overhead) the mussels may lose their byssus attachments and drop to the muddy bed where they perish. The same phenomenon has been observed in another system of hanging culture in which the mussels are weakened by the parasitic copepod *Mytilicola intestinalis.* Selection of appropriate hibernation sites, which might perhaps be found further offshore in deeper water, deep enough to keep the mussels free from the impact of waves, is an important matter.

Occasionally one observes a drop in the oxygen level in periods of prolonged calm weather. It is not yet clear why the phytoplankton in such cases does not produce enough oxygen during the daylight hours to compensate for the general oxygen consumption. It could be that the death of plankton blooms is responsible for such a phenomenon, but further research in this field is needed to gain a good understanding of what happens. In summer plankton blooms can be built up by dinoflagellates, among which are species of *Goniaulax* and *Dinophysis* which produce toxins that could render the mussels unfit for human consumption. Until now there have been no recorded cases of toxic plankton causing Black Sea mussels to be poisonous to man. If the phenomenon does occur here it might be only in the summer season, which is the close season for the local mussel industry, and thus the consumers are safeguarded.

C.2. Predators

There are not many known predators of the mussels in the Black Sea, and the rich and extensive natural beds occurring there demonstrate that predation cannot be of a serious nature. Starfish do not occur here, presumably due to the rather low salinities. Some demersal fish, especially several species of goby *(Gobius fluviatilus, G. melanostomus)* are said to devour mussel seed, but quantitative data are not available. Perhaps the most notorious mussel enemy is a newcomer, the large Japanese drill *(Rapana thomasiana)*, a gasteropod which was accidentally introduced into this area in the 1940's, and has spread rapidly. Since *Rapana* has no pelagic larvae it cannot affect mussels in hanging culture.

C.3. Parasites

No mention is made of the occurrence of the parasitic copepod *Mytilicola intestinalis*, though it is found in many places along the shores of the Mediterranean. If it does occur in the Black Sea it could perhaps explain why the meat content of the mussels from the natural beds is so poor there in comparison with other mussel areas, and why mussels from hanging culture are so much better in quality, as they would probably escape from the infestation of *Mytilicola* larvae which seek their host at the bottom of the sea. But thus far *Mytilicola* has not been found in the mussels of the Soviet coast of the Black Sea, nor in the coastal waters of Rumania and Bulgaria.

A parasite which put its stamp on the entire Black Sea mussel industry is the little trematode worm which is reponsible for the pearls found in such prohibitive numbers in the mussels from the natural beds, and which really form the main constraint in the rational exploitation and processing of this natural resource. Since the days of Sinitsin (1911) it has been known that the Black Sea mussels are affected by the metacercaria of the Gymnophallid trematode now named *Gymnophallus margaritarum*, then called *Adolescaria perla*. This gives rise to the pearls in the mussel meat. Sinitsin assumed that the sporocysts and cercariae in *Tapes* and *Cardium* are the progenitor stages of the metacercaria in *Mytilus*, but he did not find the adult stage which in all probability exists in some kind of diving waterfowl. Further research in this field might throw more light on this phenomenon and perhaps will provide a solution.

C.4. Competitors

Quite a few invertebrates live on, and between the mussels on the natural beds, but as everywhere else, their impact is greater in a hanging culture. Evidently organisms such as ascidians, bryozoans and balanids settle more easily and grow better on the shells of the mussels which are exposed to water from all sides than on mussels living on the sea bed. They compete for food and space and may interfere with the growth and fattening of the mussels, occasionally also with processing. Thus far no methods of control are contemplated or practised.

D. ECONOMIC ASPECTS

Since mussel farming is still in the experimental phase in the Black Sea it is too early to analyse its economic aspects and to give exact figures for inventory items, expendable items, personnel and wages, production, sales, etc.

It can be stated that fishing on the natural beds costs from R 60 to 90 per 1 000 kg of marketable product, and that the value of mussel meat is estimated at R 1 100/metric ton. It is furthermore interesting to note that one may find some mussels on the local markets, collected by individual fishermen. They are sold for a price varying between R 0.40 and 0.60 per 100 mussels, which corresponds to R 0.30 to 0.45/kg, a price of the same order of magnitude as encountered in other countries. The official bag-limit of 8 kg is not strictly enforced since there is no risk at all that the stocks of the natural beds will be unduly affected by the activities of individual fishermen.

It is interesting to note the market prices of some of the canned products per 250 g net weight: (1) Black Sea mussels in their juice, R 1; (2) Mussels with buckwheat, R 0.58; (3) Plov de moules à la Taurida, R 0.58; (4) Mussels with beetroot and prunes, R 1.05; (5) Dietary mussels with prunes, R 1.10.

Something could be said about the expenses incurred in building an experimental mussel farm, using wooden poles, wire and anchors, the former being used for one season only, the latter repeatedly. The PVC blocks cost R 1.24/kg, which in the present phase leads to an expenditure of about R 620 per year for making PVC plates. Such plates could, however, be used more than once. For the present experimental phase, it makes little sense to assess how much the 12-m research boat, with its 150 h.p. engine and its crew of three men, costs, and how many days it works for each of the different operations of the mussel venture with a scientist and two field technicians of AzCherNIRO on board and possibly with extra SCUBA divers; about 6 days per month are now spent working with two or three contracted divers. On commercial lines mussel farming will show a very different pattern for then one should only include the real work in the park, such as bringing out and harvesting the collectors at the most appropriate time, the construction of the parks in spring and their clearing away in autumn, the costs of the hibernation system, etc. It is the duty of the economists of AzCherNIRO to study the economic aspects of mussel farming in its probable future form in the Black Sea, and to compare that pattern with the present form of exploitation of the natural beds. On the basis of the experiments which have been carried out already they conclude tentatively that it will be possible to produce mussels by farming at about half the price of fishing.

E. THE ROLE OF THE GOVERNMENT

Governmental agencies, especially AzCherNIRO, operating under the U.S.S.R. Ministry of Fisheries and directly financed by the State, now carry the whole burden of experimental mussel farming in the Black Sea. It is the section for commercial invertebrates, under the guidance of the biologist A.I. Ivanov, which is charged with this work and it is fortunate that AzCherNIRO has also a section for Technical and Economical Research which can render assistance in this venture. A research ship of the AzCherNIRO collects hydrographic data and plankton samples, and is also used for the farming operations in the experimental parks. Since the number of experimental parks varies from year to year, the number of people engaged in this research is not always the same. In the year 1973, for instance, eight workers from the AzCherNIRO laboratory, including scientific workers and technical assistants, worked full-time on the mussel project. AzCherNIRO is paving the way towards mussel farming in the Black Sea, and when sufficient positive results have been obtained to warrant the setting up of truly commercial mussel farms, it is thought that State-owned enterprises and/or cooperative mussel farms will take over, but that AzCherNIRO will keep the leading role in the field of biological, technical and economic research and advice.

F. GENERAL IMPRESSION

The efforts to switch over from mussel fishing to mussel farming on the northern shore of the Black Sea are of considerable interest and importance. Despite the existence of extensive and rich natural mussel beds, the exploitation of these by fishing alone is not very attractive and will not lead to a large production of high quality food for human consumption. This has been explained in terms of constraints, mainly caused by the necessity of separating by hand mussels bearing small pearls from those which are free of these troublesome calcareous concretions, combined with the need to sort and grade the mussels on board. For fishing and processing mussels at sea one could certainly think of using more efficient tackle and specially-designed ships, taking advantage of the experience gained in other countries, for a seine-net vessel is certainly not a very practical ship for mussel fishing. The pearl problem is more difficult to solve, especially since too little scientific evidence is available for a good quantitative understanding of this phenomenon.

Cultivated mussels, growing faster, developing a better meat quality, and devoid of pearls, will certainly be more readily accepted on the market than the mussels from the natural beds, even though the former are processed. Therefore the efforts to create mussel farms are of considerable practical importance. An interesting aspect is that both the present fishing and the future farming will be State operations and not private enterprise. This may very well prove to be a considerable advantage. Private enterprise may be good at the exploitation of a potential resource when the method is already more or less fool-proof, leading to consistant economic results and when the government has sufficient control over the various legal aspects of such enterprises to prevent wasteful exploitation. But private enterprise brings the temptation to aim at quick profit. Therefore the temptation exists to start large-scale operations based on too little experience, which may lead to great financial loss when reverses of fortune are encountered and the whole venture is dropped, even if the venture can be considered basically sound. When State agencies lead such a development, and are not lacking in money and manpower to work on a technical scale, there is every chance that the operation will not run too fast, progress will be made step-by-step from a laboratory scale to a semi-technical scale, and only from there to a commercial scale when enough insight and experience have been gained. Since the governmental agency under consideration has the advantage of a multi-disciplinary set-up, including biology, chemistry, engineering and economics, there is every hope for ultimate success.

The problems to be solved are still serious:

(a) Spat collecting with plates held in a plastic cord is possible, but requires a lot of manual labour, and is difficult to mechanize. Harvesting half-grown mussels and marketable mussels from such collectors is also difficult to carry out on a large scale. It might be profitable to experiment with coco-fibre rope which in France proved to be so very efficient as a collector that its use had to be restricted.

(b) In the experiments so far carried out the mussels are kept on the original collectors until they have reached the marketable size. In all other countries farming is carried out in two steps, production of half-grown mussels, followed by growth and fattening up to consumption size. This involves transplantation, either to parks where they grow and fatten more rapidly, or to new strings hung from rafts, poles or racks. In several countries cylindrical nylon netting has proved to be a very efficient material in which to transfer mussels to new growing sites. It might be worthwhile to try this method in the Black Sea too.

(c) If coco-fibre ropes were used to collect the seed, or sausage-like nets filled with half-grown mussels put out, it would be interesting to try to wrap these, with the help of SCUBA divers, spirally around poles placed deep enough to escape from ice floe damage in winter.

(d) Hibernation is still a serious problem in hanging culture in areas where ice and gales may threaten the mussels in winter. The present metal frames, which are set up in mid-winter are interesting, but it might be an advantage to test hibernation in deeper sites, well under the layer affected by ice and waves, after ascertaining that the water will be of a sufficiently high quality throughout winter.

(e) If hibernation is found not to be technically and economically feasible, harvesting the young mussels born the same year might be considered. When processed, the size of the mussels is of less importance than when the product is marketed fresh. An official reduction of the minimum legal size for farmed mussels is within the realms of possibility if such is found to be of economic interest.

(f) Though sea bed culture as it is practised in The Netherlands does not seem possible at first sight, primarily because of the danger that pearls will grow in the farmed mussels, it would be interesting to carry out some experiments in this field on well-selected non-shifting hard grounds, in places where the water carries enough phytoplankton, but where mussel seed does not settle naturally.

G. BIBLIOGRAPHY

Berenbeim, D.Y., 1957. New features in the hydrological picture of the Sea of Azov and the Kerch Strait. Priroda (Moscow), 2 (in Russian).

Borisjak, A., 1905. Pelecypods from the Black Sea plankton. Izv. Akad. Nauk SSSR, 22, Ser. 5 (in Russian).

Caspers, H., 1957. Black Sea and Sea of Azov. Treatise on Marine Ecology and Paleoecology. Geol. Soc. Am. Mem., 67: 801—899.

Ivanov, A.I., 1963. The mussels from the Black Sea. Rybn. Khoz. (Moscow), 11: 23—27 (in Russian).

Ivanov, A.I., 1965. The mussel stocks in the northwestern Black Sea. Rybn. Khoz. (Moscow), 10: 15—18 (in Russian).

Ivanov, A.I., 1967. Growth of the Black Sea mussels on the Odessa Bank. Gidrobiol. Zh., 3 (2) (in Russian).

Ivanov, A.I., 1969. The distribution of mussels containing calcareous "pearls" in the principal exploitation districts of the northwestern Black Sea. Tr. AzCherNIRO No. 26, Pishevaja Promyshlennost Publishers, Moscow (in Russian).

Ivanov, A.I., 1971. Preliminary results of cultural rearing of mussels (*Mytilus galloprovincialis* Lmk) in the Bay of Kerch and in some Black Sea areas. Okeanologiya, 11 (5): 889—898 (in Russian).

Ivanov, A.I., 1971. The first Soviet experiment of culture of mussels. Rybn. Khoz. (Moscow), 9: 18—19 (in Russian).

Korobkina, G.S., Dannilova, E.N., Kalinina, N.W. and Leonora, T.A., 1965. The food value of Black Sea mussels. Rybn. Khoz. (Moscow), 12: 57—59 (in Russian).

Potakov, G.I., Rabinovich, V.I., Tatunov, B.N. and Trofimov, V.P., 1965. Improved equipment for mussel production by the medium-size seiners. Rybn. Khoz. (Moscow), 12: 48—51 (in Russian).

Sinitsin, D.F., 1911. The parthenogenetic generation of the Trematodes and their descendants in the Black Sea Molluscs. Zap. Imp. Akad. Nauk St. Petersb. Cl. Phys. Math., 30: 1—127 (in Russian).

Trofirnov, V.P., 1970. Mechanized Production of Marine Invertebrates. Pishevaya Promyshlennost Publishers, Moscow (in Russian).

Woroniev, W.P., 1938. The Mussels of the Black Sea. Tr. AzCherNIRO, No. 2, Goskrymizdat, Simferopol (in Russian).

Zachvatkina, K.A., 1963. The phenology of the larvae of the bivalve molluscs of the Bight of Sebastopol. Tr. Sevast. Biol. Stn., 16. Izd. A.W. USSR, Kiev (in Russian).

Zenkevich, L.A., 1947. Fauna and Biological Productivity of the Seas. Vol. II. Seas of the U.S.S.R. Sovyetskaya Nauka, Moscow, 588 pp. (in Russian).

Zernov, S.A., 1913. Contribution to the knowledge of the life in the Black Sea. Zap. Imp. Akad. Nauk St. Petersb. Cl. Phys. Math., 32: 1—299 (in Russian).

Chapter 8

FARMING THE GREEN MUSSEL *(MYTILUS SMARAGDINUS)* IN BACOOR BAY, THE PHILIPPINES

Example
Tahong farm of Mr Magtangol Alagar, Barrio Siniguelasan Bacoor, Cavite, Luzon, The Philippines.
Period of observation: 1971—1972.

A. BACKGROUND

A.1. General principles

Farms for molluscan shellfish are predominantly located in the temperate zone of the northern hemisphere, in some subtropical areas, and but rarely in the tropics. The Philippines form a notable exception, for here one finds oysters farms in many places, and mussel farming has been taken up quite recently. The molluscan shellfish, which is rich in protein, is a very welcome item of diet for the dense populations of these islands.

Tropical conditions require a special adaptation of farming techniques, since exposure to the tropical sun endangers the life of the shellfish. Furthermore, transportation and marketing of live shellfish is not an easy matter in the tropics.

The green mussel, locally called tahong, is farmed in subtidal sheltered sites. To collect the mussel seed extensive bamboo structures are used in places where the sea bed is rather muddy. Where acorn barnacles (Balanids) have settled in large numbers on the bamboo one may hope for a good settlement of mussels, well below the low water mark. The mussel farmers in Bacoor Bay do not practise transplanting or thinning out the mussels, but keep them on the bamboo supports until they reach the marketable size, roughly 6 months after settling. The mussels are harvested by divers, who pull the poles carrying the mussels out of the mud, and strip them clean on board a boat. Having been well-graded and cleaned, the mussels are rapidly brought to the market, where they are sold alive.

Several systems of park construction were devised on the basis of experience gained in oyster farming but it was observed in due course, that an open construction is preferable because of greater strength, and better growth and fattening of the mussels. Formerly the so-called "wigwam" method was used on a large scale, in which eight long bamboo poles were placed in a circle of 4 m diameter around a central pole, and nailed together with short horizontal bamboo braces just above the low water mark. Slender bamboo poles were then placed inside the frame, using the central pole as a pivot. Experience has shown that the "wigwam" method may easily lead to overcrowding and poor growth, and moreover, that breakage of one pole may easily lead to a collapse of the whole structure. Therefore virtually all the Bacoor Bay mussel farmers, including the firm chosen as an example, nowadays practise the "stake" system, characterized by rectangular bamboo structures.

Since mussel farming Philippine-style requires only two labour-intensive operations, i.e., building the bamboo structure and harvesting the crop, this underwater work is carried out mostly by divers hired for the occasion.

Most of the farmers work on a small area of 0.25 ha. Such an area may yield per year about 2 500 gallons (about 6 500 kg) of mussels in the shell.

A.2. Biology of the species

The Bacoor Bay green mussel, *Mytilus smaragdinus*, locally called "tahong", resembles so closely the mussels of the same genus grown in Europe, that its biological features need not be described in any detail here.

Typical of this mussel is the green hue of the shell's outer surface. Like other mussels, they often grow together in clusters, and such a cluster looks like a beautiful green flower, the radially-placed individual mussels representing the petals.

The preferred marketable size is 4 to 6 cm shell-length, attained within 6 months after settling, but the mussels of this species can reach a much larger size, up to 25 cm.

Like other mussels, the tahong is herbivorous, feeding mainly on phytoplankton and detritus particles, filtered from the water in which they are bathed. Spawning may take place in any month of the year, which is not surprising under tropical conditions. The pelagic larvae need a suitable substratum to settle on. Newly-placed smooth bamboo poles do not attract the larvae but when they become covered with barnacles, one may expect a good set of mussel seed on them. In due course the mussels will completely oust the barnacles, adhering firmly to the bamboo support with their byssus threads.

A.3. Geographical situation

The 0.25-ha mussel farm of Mr Magtangol Alagar is to be found in Bacoor Bay, the section of Manila Bay due east of the Cavite Peninsula at the exact geographical position 14° 28′ 40″ N, 120° 57′ 20″ E, some 14 km south-southwest of the city of Manila (see Chart).

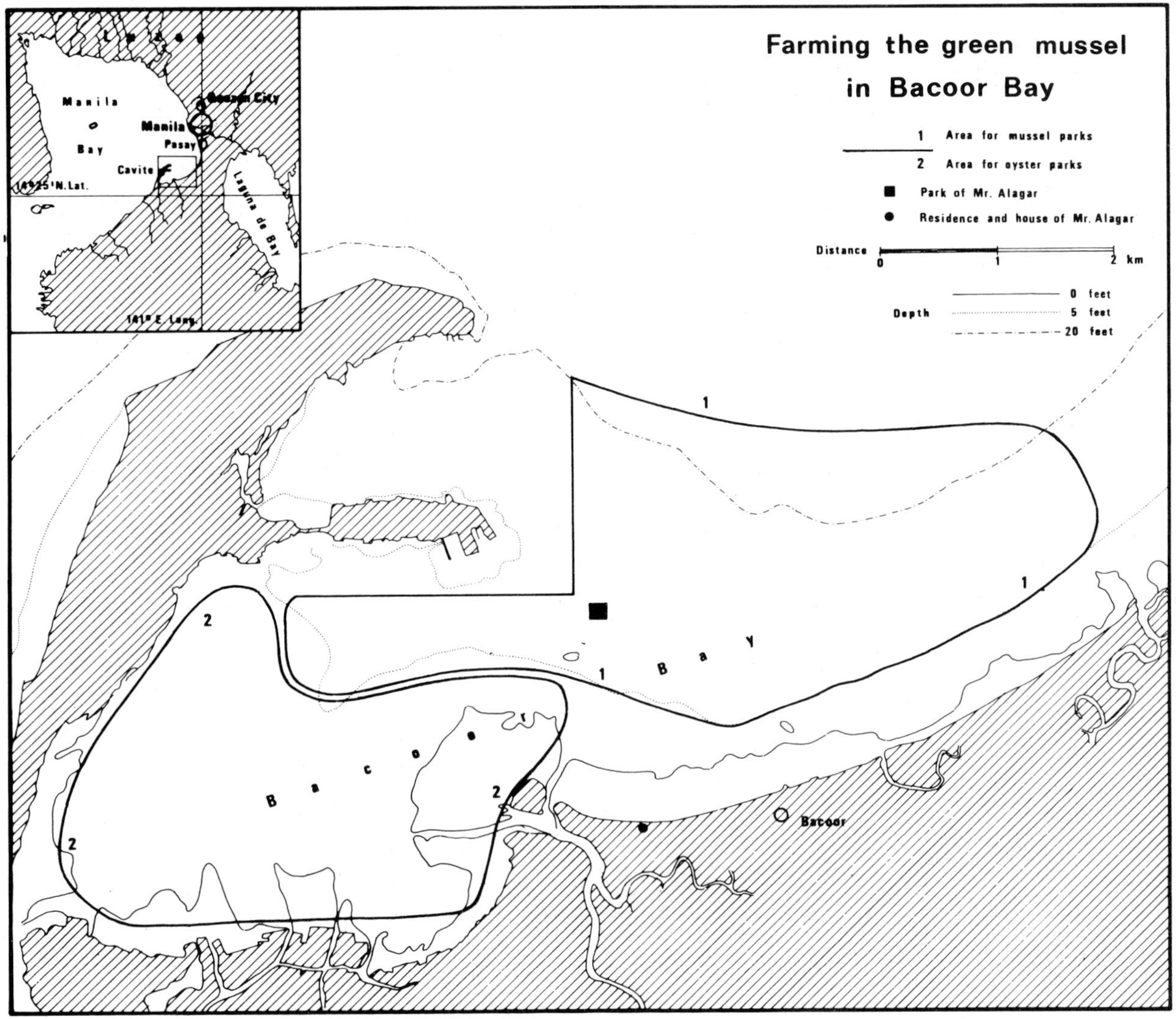
Farming the green mussel
in Bacoor Bay
1 Area for mussel parks
2 Area for oyster parks
Park of Mr. Alagar
Residence and house of Mr. Alagar
Distance
0
1
2 km
Depth
0 feet
5 feet
20 feet
Manila Bay
Manila
City
Pasay
Cavite
14°25'N. Lat.
141° E. Long.
Laguna de Bay
Bacoor Bay
Bacoor
1
2

A.4. Hydrographic pattern

Depth. The mussel parks of Bacoor Bay are all situated well below the low water line, at depths from 2 to 7 m at low tide. The mussel park under consideration is about 4 m deep.

Tides. The tides in Manila Bay are predominantly of the diurnal type and show a range of 1.20 m at spring tide and 0.60 at neap tide. Hence the tidal currents are not very strong, rarely exceeding 0.5 knot.

Waves. Bacoor Bay is well-sheltered. Only when strong winds blow from the north do the waves show an appreciable height. But even then, the parks are hardly affected, since they are fully submerged, even at low tide.

Water temperature. The water temperature varies only slightly in the course of the year in this tropical area. Surface temperatures show a winter minimum of 26 to 28°C and maximum values up to 31 or 32°C in mid-summer.

Salinity. The salinity in Bacoor Bay fluctuates usually between 27 and 33‰ but may drop as low as 20‰ in the surface waters in periods of heavy rainfall. Therefore the salinity figures for September are often low. Hurricanes followed by floods can severely disturb the general pattern, as was experienced here in August 1972.

Plankton. The water of Bacoor Bay is usually quite clear with a distinct greenish hue. There are some data available on the total volume of net-plankton in the inshore part of Manila Bay, both for the dry season (May 1970) and for the wet season (August 1970). They show for Bacoor Bay about 4 cm^3 of plankton per m^3 in the dry season and 2 cm^3 in the wet season.

Pollution. Usually the water contains about 5 mg/l of oxygen in this area, but when the wind and currents come from the north, pollution caused by the sewage discharged by the city of Manila may lead to low oxygen concentrations in Bacoor Bay. When this happens at neap tide, there is an increased risk of mortality among the mussels.

Sea bed conditions. Where the tahong farms are constructed the sea bed consists of a sandy mud.

A.5. Legal aspects

Mr Magtangol Alagar is the proprietor of his mussel farm; it is a case of private enterprise. Though responsibility for the mussel parks of Bacoor Bay falls officially under the office of the Philippines Fisheries Commission, it is in practice the Municipal Government which regulates the use of the parks. The rule applied now is that one farmer cannot own more than 0.25 ha of mussel park, unless he has already been farming a greater area (up to 2 ha) under the previous regulation. The farm under consideration measures 0.25 ha (50 × 50 m) for which the lessee has to pay a municipal tax of P 18 per year.

There are no guards on the parks, and police patrols are infrequent. Nevertheless, theft and poaching are rare. The fully submerged parks cannot be reached on foot and are not easily found at night in the dark. The farmers, all members of an association, survey each other's parks by mutual agreement. They will report immediately when thieves or poachers from elsewhere are seen on someone's plot. Since the parks are small and numerous there are always quite a few farmers on them during daylight hours.

B. SEQUENCE OF OPERATIONS

B.1. Building a mussel park

Our mussel farmer, like all his colleagues in this area, uses the stake or "tulos" method to farm his mussels. Since his park is of medium depth he can use the variety of bamboo locally called "buho" *(Schizostathyum lumapao)* from which he cuts stakes 3 to 5 m long and 3 to 4 cm in diameter at the butt end, tapering to 2 cm at the thinner end. Farmers operating at deeper sites need the longer, more expensive stakes cut from the true bamboo *(Bambusa arundinacea).*

The stakes, brought to the park by boat, should be placed in rows in the muddy sea bed, with 40 to 50 cm between adjacent poles. On a 0.25 ha park about 5 000 poles are systematically placed this way, leaving some lanes open for navigation.

The farmer hires a group of divers to place the stakes. The divers are equipped with wooden-framed glass goggles and wooden flippers. One vigorous stroke with both feet gives sufficient force to drive a pole deep enough into the mud, and a diver can stake two or three poles per dive. Five men can stake 1 000 buho poles in 2 to 3 h. Placing stakes on the park can be done at any time of the year, but is in preference carried out from April to October. Directly after harvesting the marketable mussels the stakes may be placed out anew. The old stakes may be used again if they are still strong enough, but "buho" poles do not last longer than 1 year.

Farmers working with longer poles in the deeper parks have more work to do, since they have to interconnect the vertical poles with long and short horizontal sticks of bamboo respectively called "bilas" and "barals". Though it is the usual practice to hire a group of divers to construct a park, there are farmers who do all this work themselves, thus saving on wages.

B.2. Inspecting the mussel seed

When the stakes are placed on the park, the settling of mussel seed is left to nature. The farmer's experience is that poles placed in the period April—October will readily collect a set of barnacles ("taliptips"), and it is on the shells of the barnacles that mussel seed will readily settle. It is not known whether the barnacles themselves allure the pelagic mussel larvae or small hydroids growing on the calcareous shells of the barnacles, as is the experience in Europe, where the mussel larvae evidently like to grab some filamentous object for their initial settlement. Thereafter they can move about with their agile foot and fasten themselves with byssus threads to any substratum which pleases them.

1 or 2 months after placing the poles, the farmer expects to find young mussels on them, usually most abundantly some distance above the sea bed and not too close to the surface of the water. Some 2 000 to 3 000 seedlings per m is said to be a reasonable number for the middle section of the pole. During this phase the farmer inspects his stakes regularly by diving, though there is little he can do to protect his young mussels.

B.3. Growing mussels to the marketable size

It takes the mussels about 6 months to reach the marketable size of 4 to 6 cm shell-length. At this size they are very tender and palatable, and produce more meat per gallon than would a gallon tin filled with larger live mussels; hence their higher price per gallon.

The mussel farmers do not thin out their mussels during the growing period, nor transplant them to other parks where growth and fattening might be faster. They have observed, however, that growth and fattening is poor on the shallowest parks close to the shoreline, and that the most rapid growth occurs well offshore on deeper parks. Mussels from the latter parks have the disadvantage of a poorer keeping quality and will die sooner after harvesting than those grown on poles at medium depth.

It is up to the farmer to decide when to harvest his crop, taking into account the size of his mussels and the demand of the market. Occasionally the mussels are kept in the park as long as a year, after which some may have reached a shell-length of 10 cm.

B.4. Harvesting the marketable mussels

There is no fixed period of the year for harvesting and marketing mussels in The Philippines. Since spawning may take place at any time of year there is no need for a close season to cover a period when the mussels are too lean to be marketed. In fact, mussels are consumed in every physiological condition. When the gonads are filled with eggs the mussels are a reddish orange colour. The farmer need not be concerned whether the mussels he harvests are in the spawning or resting condition.

The decision to harvest is made primarily when a buyer orders a given quantity of mussels for a given day. Then the farmer hires divers for the preceding day. The divers, equipped with goggles and flippers, go to the park with the farmer's boat, and pull out the stakes, which are then loaded onto the boat. Other hired workers strip the stakes, making use of an iron bar, sharpened at one end; they grade the mussels according to size and wash them carefully to free them from possible epibionts and to eliminate the adhering mud. All this work is carried out on board. Small mussels and mussel seed found among the marketable specimens are not used as primary material for farming, but are discarded.

Usually the farmer hires a team of five divers when he has received a substantial order. Such team can collect about 500 gallons of mussels in 2 to 3 h when the poles are well-laden. For a smaller consignment the farmer may decide to go and get the mussels himself.

B.5. Marketing the mussels

The quality of mussels is measured by the gallon, using a gallon tin, and the buyer waits at the wharf to receive them, thoroughly checking the grading, the quality and the cleanliness, before accepting the consignment.

The mussels are usually packed in jute sacks or in baskets made of bamboo or rattan for quick transportation to metropolitan Manila during the night or in the early morning. Wholesalers distribute them to market retailers.

Usually the mussels reach the market within 5 h of harvesting from the parks. Without refrigeration they can be kept alive for, at most 24 h.

The Philippine mussels are fortunately not eaten raw, since they are grown in an area occasionally affected by domestic sewage. They are always cooked to be used in one of a variety of dishes such as "tahong Q" and "adobo" [1], and are often minced to be an ingredient in various dishes such as "relyeno" and "embodito", which require enrichment with animal protein.

C. FARMING RISKS

C.1. Hydrographic conditions

During periods of heavy rain the salinity may drop considerably, occasionally so much that the life of the mussel becomes endangered.

Another unfavourable factor is the wind from the north which may send heavily-polluted water to the parks. When this happens at a neap tide, when flushing is poor, low oxygen values may occur, killing the mussels.

Typhoons are to be feared more than anything else. The typhoon season lasts from July to October, during the northeastern monsoon, and it is impossible to predict where a typhoon will strike. August 1972 brought disaster to the Bacoor Bay mussel farmers; a typhoon hit their parks and thereafter heavy floods, lasting for weeks, caused the salinity to drop so much that all the mussels in the area died. Strangely enough, oysters settled profusely in some of the devastated mussel parks, that of Mr Alagar included, but early in 1973 some mussel seed was found again.

C.2. Predators

The mussel farmers of Bacoor Bay do not complain about predation upon their mussels. The mussels are quite sturdy and evidently there is generally enough mussel seed on the bamboo poles to permit some losses.

C.3. Competitors

Boring organisms such as shipworms may occasionally gather in such great numbers in the poles that these become brittle and break when the load of mussels gets too great. This only happens when the shipworms settle earlier than the mussels. As soon as the poles are covered with mussels shipworms can no longer find a place to settle.

The Bacoor Bay mussel farmers do not complain about epibionts settling on the shells of their mussels, competing for space and food. The mussels are freed of such organisms before they are marketed.

D. ECONOMIC ASPECTS

D.1. Rents

For a park of 0.25 ha a municipal tax of P 18 per year has to be paid.

D.2. Inventory items

The mussel farmers of Bacoor Bay usually have their own boat, called "banca", a simple dug-out canoe, measuring from 7 to 10 m in length, with a wooden framework covered with waterproof plywood to increase the height, and a width from 40 to 60 cm. Such a boat is usually equipped with an inboard gasoline engine of 1.5 to 12 h.p.; it has quite a long life and is inexpensive. For a boat of 10 × 0.5 m one should expect to pay P 2 000 whereas a 9 h.p. inboard motor of the brands "Briggs and Stratton" or "Wisconsin" would also cost about P 2 000. For mooring the boat one needs some wharf space. The only tools one needs are some scrapers, a knife and rattan baskets.

D.3. Expendable items

Every year the mussel farmer has to purchase a fair number of bamboo stakes. For a farm like the one considered here, working with buho, one should expect to need 5 000 stakes for 0.25 ha, since each stake lasts about 1 year. The buho stakes, 4 to 6 m long, are purchased from dealers at a price of P. 0.20 each. Farmers operating on deeper plots have to buy longer bamboo stakes (8 to 10 m long, 10 cm in diameter at the butt end) at a higher price (P 1.20 each) but they will last up to 2 years.

The farmer needs gasoline for the engine of his boat. Since the distance from the wharf to the park is only about 3 km and frequent visits are only necessary at the time the poles are erected and when the mussels are harvested, this item does not weigh heavily on the farmer's budget.

D.4. Personnel and wages

Our mussel farmer has no personnel in permanent employment. When the stakes have to be put out in the park he contracts a group of divers, usually five. Such a group can stake 1 000 stakes in 2 to 3 h. The wage at the time the farm was visited was P 50/1 000 stakes. The entire 0.25-ha park contains about 5 000 stakes.

At the time of harvesting, the farmer again hires a group of workers for diving and pulling out the stakes covered with mussels, for scraping the mussels off, and for grading and cleaning them on board. Five male workers can gather 500 gallons of mussels in 2 to 3 h when the poles are well-laden. They are paid P 0.30—0.40/gallon.

D.5. Production and sales

The production may fluctuate rather widely according to the length of the poles used, and the quantity of mussels per metre of pole. Poles of the buho type, as used on the farm under consideration, may be found to carry 2 gallons of mussels; 1 gallon of mussels in the shell weighs about 2.5 kg. A total production of 5 000 gallons per year per 0.25 ha park is considered as normal.

The price paid for the mussels varied during the time of observation from P 1.50 to P 2/gallon. This is the price the farmer gets; the consumer will have to pay P 0.50 to 1.00 more.

E. GOVERNMENTAL ASSISTANCE

The mussel parks of Bacoor Bay consist of small plots, each measuring 0.25 ha. It is evidently the policy of the municipality to allow a great number of small farmers to make a living. Intensive farming on small plots leads in all probability to a greater total production than more extensive farming on larger plots. The total area of suitable sheltered ground is limited in

Bacoor Bay to about 600 ha. The rent to be paid by the lessees is low: P 18/ 0.25 ha/year.

The Philippines Fisheries Commission takes an active interest in shellfish farming and it was, in fact, the Commission's fisheries biologists who carried out the first experimental work on developing methods of mussel farming, in a period when the oyster farmers still considered the mussel as a sheer nuisance, competing for space and food with their crop. Investigations on various aspects of the biology and farming of oysters and mussels still form a substantial part of the research programme of the Commission.

F. GENERAL IMPRESSION

It is a good thing when a densely-populated country stimulates the rational exploitation of its natural resources in an effort to produce more food rich in animal protein and to create a living for a fair number of its inhabitants. Based on the experience of oyster farmers, a mussel industry has in recent years been created, assisted by the biologists of the Philippines Fisheries Commission.

Science fiction writers would have us believe that there will one day be underwater gardens in which frogmen tend to a crop, protecting it against predators and competitors, but here in The Philippines this has already become a reality; a valuable crop of sea food is produced subtidally and divers carry out most of the work involved!

Surely it is possible to find more localities suitable for mussel farming in The Philippine archipelago and it is feasible to transport sections of bamboo carrying live mussels, wrapped in wet cloth, to sites where one wishes to introduce this species, but before doing this on a large scale it would be wise to study the various farming risks in greater detail. Of these should be mentioned the occasional occurrence of low salinities, and the relationship between this and typhoons. Another factor of importance is quick transportation to the market, since mussels cannot survive out of water for more than 24 h without refrigeration. Therefore it is best to establish the farms close to a big city such as Manila, where the product can easily be sold, but on the other hand the proximity of a city involves the danger of pollution with domestic sewage, which may lead to conditions which endanger the mussels on the farm, especially at neap tides. Mussel farming in more isolated places, far from sources of pollution, meets with the difficulty of finding enough workers and of bringing the product alive to a large market within a few hours. Locally the mussel meat is preserved in brine, but the fresh product is greatly preferred by the general public.

The Philippines mussel farming venture is certainly very interesting. The amount of capital required to start a 0.25 ha farm is minimal, of the order of a few thousand Pesos, so that it is not too difficult to make a modest living by farming mussels. The required materials, different varieties of bamboo, are readily available at a low price and occasional workers, to be contracted as divers for a few days, are easily found in the densely-populated sections of the Philippines. It must be possible to produce and market a considerable tonnage of green mussels in this part of the world, and contribute materially to the wholesomeness of the diet of the local population, but it seems too early yet to speak, as Obusan and Urbano do, about "food for the millions".

G. TECHNICAL ASPECTS

G.1. Tahong recipes

(a) Tahong Q

Market list: 150 large pieces of tahong meat
1 clove of garlic
2 cups pineapple juice
½ cup soy sauce
2 teaspoons food seasoning (vet-sin)
1 spoon ground black pepper
salt
50 barbecue skewers

Procedure: Shell tahong. Marinate in pineapple juice, soy sauce, crushed garlic, seasoning, salt and black pepper for 4 h. Remove from mixture. Stake three pieces on each skewer. Fry with skewer in deep fat till golden brown. Serve hot or cold with ketchup. Makes 50 tahong Q.

(b) Tahong Adobo

Market list: 2 cups shelled tahong
1 medium size onion minced
5 cloves of garlic
1 cup vinegar
1 teaspoon black pepper
soy sauce to taste

Procedure: Shell tahong and set aside to dry. Bring vinegar, soy sauce, crushed garlic, onions, and pepper to boil. Add shelled tahong and cook for 2 min. Stir occasionally. Serve hot. Makes 4 servings.

(c) Tahong Rebosado

Market list: 30 large pieces Tahong (shelled)
2 eggs (beaten)
3 spoons flour
½ teaspoon salt
2 cups oil

Procedure: Mix flour, baking powder, salt, pepper and beaten egg. Dip each tahong into mixture before frying in deep oil until golden brown. Serve with ketchup.

H. BIBLIOGRAPHY

Escritor, G.L., 1966. Tahong. Philippine Fisheries Yearbook, Manila, 36—45.

Escritor, G.L. and Padua, P., 1961. The culture of *Mytilus smaragdinus*, a Bay mussel of Bacoor Bay, Cavite. Diliman, Quezon City, P.I.

Faustino, L.A., 1928. Summary of Philippines marine and freshwater molluscs. Bureau of Printing, Manila, pp. 35—39.

Obusan, R.A. and Urbano, E.E., 1968. Tahong, Food for the Millions. Indo-Pac. Fish. Counc. 13th Sess. Brisbane, C. 68/Tech.29. Occasional paper 69/13, 25 pp.

Chapter 9

FARMING LITTLE-NECK CLAMS ON THE ALGARVIAN COAST, PORTUGAL

Example
Mr Francisco Reis Bom, clam farmer at Fuseta, Algarve, Portugal.
Period of observation: 1971—1972.

A. BACKGROUND

A.1. General principles

Sea food is an important item of the Portuguese diet. Fish ranks first, but in addition both molluscan and crustacean shellfish are highly appreciated. It is peculiar however, that the so-called Portuguese oyster, *Crassostrea angulata*, of which rich natural beds occur in the Sado and Tagus estuaries, is not appreciated by the Portuguese consumer. The general aversion to raw shellfish can be explained in terms of a lasting anxiety caused by violent typhoid epidemics early in the century. If one sees people at low tide collecting shellfish on the Sado beds, you may be certain that they are picking edible gastropods and leaving the oysters alone. Portuguese oysters are hardly eaten at all in Portugal, but are collected and graded as oysters used for re-laying and exported to France, Spain, England and The Netherlands. They constitute therefore a valuable natural resource, but are not farmed. Neither is the flat oyster *(Ostrea edulis)* farmed in Portugal, though the existence of a natural bed of this species in the lagoon of Aveiro, indicates that farming might be possible there. Scallops *(Pectenidae)* and cockles *(Cardium edule)* are fished and marketed on an appreciable scale, but by far the most popular molluscan shellfish is the "amêijoa", the little-neck clam *Venerupis decussata* (L.). It is this species which is farmed in the coastal waters of Algarve in southern Portugal. On many sites along the Portuguese coast activities are confined to collecting the clams on the natural beds, either by digging them out on the intertidal beds or, on deeper grounds, by fishing them with the aid of a long-shafted rake carrying a net, the so-called "berbigoeira" (which is also used for fishing the berbigão: *Cardium edule*), but on the south coast of Algarve, a style of farming has been developed. Human intervention does not go as far as in the farming of oysters and mussels in other countries, but still one can speak of farming, perhaps not of "semi-farming" but rather of "quarter-farming", if one wishes to describe it more precisely. The farmers operate on intertidal parks. The young clams are collected by hand from the natural beds in the intertidal zone, where the clams are usually sparsely distributed. The farmer buys them from the fishermen, or participates in collecting the seed clams himself, and plants the young clams densely on his parks. This is not done to offer the clams better feeding conditions than those on the natural beds, for the parks are also located on the intertidal grounds, but rather to give them a chance to grow to a larger size. There are so many collectors that all the clams would otherwise be brought to the market as soon as they reached the minimum legal size, and some before. Now the clams are kept in the parks for several years to grow to an appreciable size. Predators and competitors are controlled as far as is feasible during this period. The advantage of this style of farming is that starting from a given number of young clams produced by nature, a far greater tonnage of clams, worth a higher price per unit of weight, may be harvested compared with direct harvesting of market-

Fig. 87. In deeper water, as here in the Ría de Portimão, marketable clams are fished with long-shafted rakes to which a collecting net is attached. (Photo H. Vilela.)

Fig. 88. Sculpture at Cambados, Spain, showing little-neck fisher at work. (Photo P. Korringa.)

able clams from the natural beds. The clams are highly appreciated on the Portuguese market and many restaurants take pride in serving them according to various styles of cooking, but a greater tonnage is exported alive, especially to Spain and France.

A.2. Biology of the species

Several species of little-neck clams occur in the coastal waters of Portugal, but *Venerupis decussata* (L.), locally called "Amêijoa boa" (or -"cristã") is by far the most important, both in tonnage and in price per kg. The species *V. aurea* (Gmelin) called "Amêijoa bicuda" (or -"de cão") and *V. pullastra* (Montagu), called "Amêijoa macha" are also of commercial importance, although they are sold at a much lower price, whereas *V. rhomboides* (Pennant) is of rather rare occurrence here.

Little-neck clams, also called carpet shells, are rather thick-shelled equivalve bivalves belonging to the Veneracea (Fig. 89). They use their foot to dig themselves so deep into the sediments that they disappear completely from sight. The extensible siphons protrude upwards when the sediments are covered with sea water, enabling them to take in appreciable quantities of water from which they collect both food and oxygen. The name "little-neck clam" suggests immediately that the siphons of these clams are not very long, in fact not nearly so long as in *Mya*, *Tellina*, or *Scrobicularia*. Therefore they do not dig very deep into the sediments; adult specimens are usually found at a depth of 5 to 10 cm, smaller specimens less deep. Though they can crawl out of the sediments to dig in again elsewhere, they do not migrate extensively, at least not after a certain age, so that the farmer can be sure that his specimens will not leave his parks.

The clams filter small particulate food from the water they take in, both detritus particles and small algae, such as diatoms, but since ingestion does

Fig. 89. Little-neck clams, stored under water. (Photo P. Korringa.)

not necessarily mean digestion it is hard to say which constituent of the food is of the greatest nutritional value for them.

The preferred habitat is a muddy sand, a rather hard sea bed in which the human foot does not easily sink. The clams are especially numerous where this type of sediment occurs in sheltered places in the intertidal zone, but locally one finds enough clams in deeper sites to make it profitable to collect them by fishing with the long-shafted rake-net. Those occurring at a depth from 4 to 12 m cannot be reached by the fishermen.

The clams are gonochorists, producing either eggs or sperm after they have grown to a shell-length of about 2 cm. Reproduction takes place in the summer season; maturation of the gonads is observed on the Algarvian coast from May to July, whereas most of the spawning takes place in August. The larvae live a pelagic life until their tiny shells measure 0.7 to 0.8 mm in length. From that size upwards they can be found in the superficial layer of the sediments on the natural beds in the intertidal zone. Observations by Vilela (1950) indicate that newly settled specimens begin to make their appearance in the month of September, but that the highest densities are to be found several months later. Evidently settling continues in October and November, possibly even later. The young clams grow steadily since water temperatures do not drop very much here in winter, and the clams reach a shell-length of 10 to 12 mm 3 to 4 months after settling, and 20—22 mm after completing 1 year of benthonic life. Though it is difficult to assess the exact age of older clams, the experience gained by the local clam farmers indicates that it takes several years before they reach their marketable size. Most of the clams are brought to the market at a size of 3—4.5 cm shell-length, but occasionally one may find older and larger specimens, up to 8 cm.

A.3. Geographical situation

The coastal stretch between Faro and Tavira is especially suited to clam farming and the two parks of Mr Francisco Reis Bom are to be found in the very heart of this area, respectively 3 and 5 km northeast of the town of Fuseta. Fuseta, situated 37° 02′ N and 7° 45′ E is where the company has its headquarters (see Chart).

A.4. Hydrographic pattern

On the south coast of Algarve clam farming is practised in the sheltered areas between the mainland and the cordon of characteristic long low sandy islands, 1 to 5 km offshore. These islands offer sufficient shelter to prevent wholesale shifting of the sediments during gales. The area behind the islands consists of extensive intertidal grounds with an intricate pattern of channels and gullies. The passages in between the islands allow free entrance of sea water with the tides.

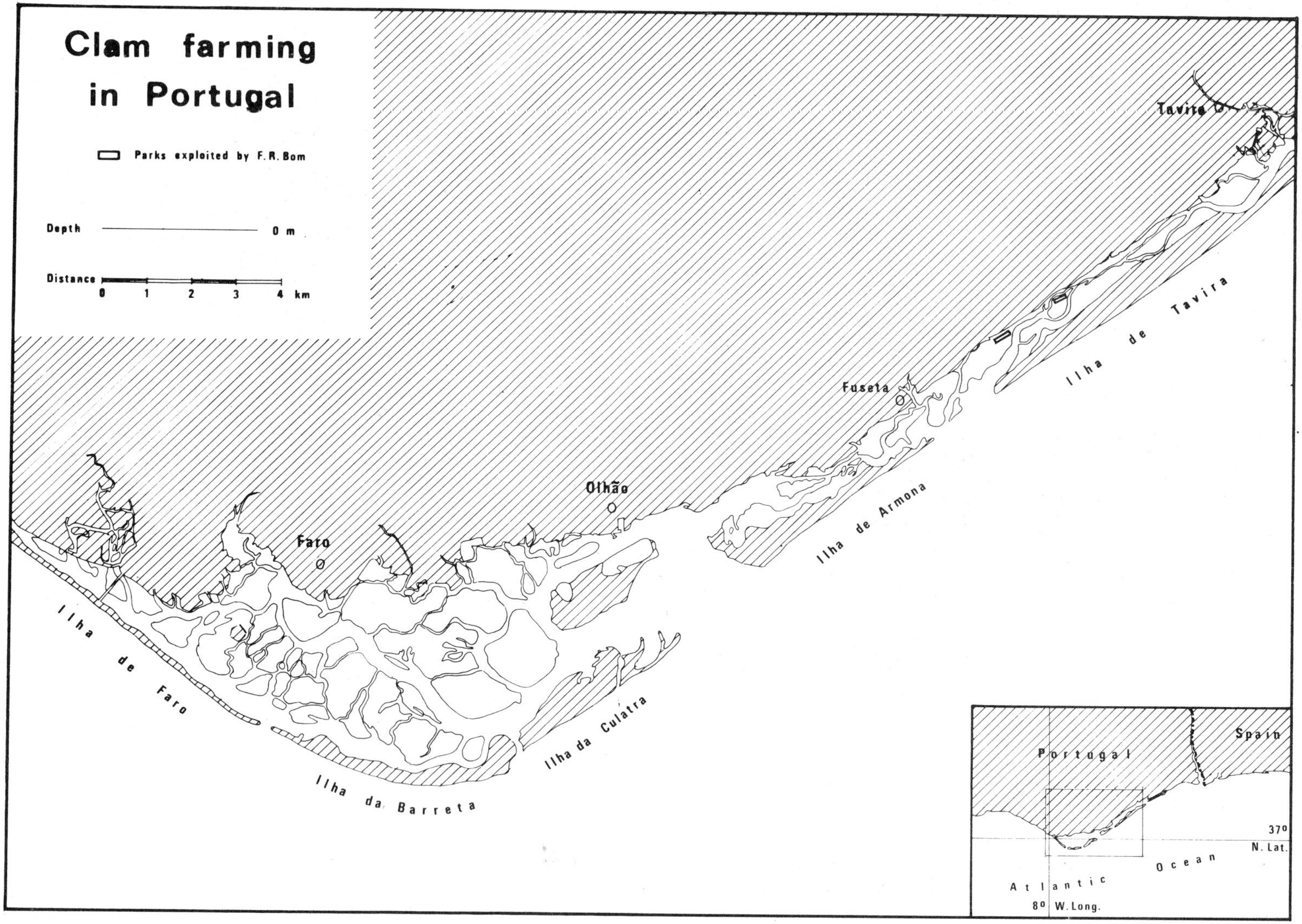
Clam farming
in Portugal
Parks exploited by F.R.Bom
Depth
0 m
Distance
0
1
2
3
4 km
Tavira
Fuseta
Olhão
Faro
Ilha de Tavira
Ilha de Armona
Ilha da Culatra
Ilha da Barreta
Ilha de Faro
Spain
Portugal
37°
N. Lat.
Atlantic Ocean
8° W. Long.

Depths. Both the natural beds and the clam parks are to be found in the intertidal zone. The best grounds are not necessarily those which are exposed at spring tide only, since the greatest numbers of clams are often found at a somewhat higher level. Sites which are exposed for many hours at a stretch and where temperatures may rise alarmingly at low tide, bear no clams.

Tides. The tidal range varies here between 2.40 m at spring tides to 1.30 m at neap tides; the tides are attended by currents of medium velocity.

Water temperature. Water temperatures in the open Atlantic Ocean near Portugal's south coast range from minimum values of 12 to 14°C in mid-winter to maximum values of 15 to 20°C in summer. Water in this temperature range will penetrate through the passages with the tide, but behind the islands water temperatures will be several degrees higher in summer (20 to 24°C) and somewhat lower in winter (down to 10° C). In the summer season, where the tidal grounds are exposed in the middle of the day, the grounds may absorb so much heat that the first flood passing over them will show very high water temperatures for a short while. It is this temperature shock which serves as a triggering mechanism in the spawning of the clams.

Salinity. The salinity of the open ocean water fluctuates between 35 and 36.8‰, but occasional rain storms may lead to a serious reduction of the salinity in the area behind the islands. Salinities of 30‰ can be measured there, and closely inshore one may even find values down to 20‰ persisting for some time. Generally speaking, the climate is rather dry in this area.

Pollution. Most of the water bathing the clam beds and clam parks is quite clean, but in the proximity of towns sewers may cause localized pollution. Government officials supervise the area every year and will not give permission to create a clam park in grossly-polluted sections. It should be taken into account, however, that the clams are always boiled or fried before they are served in Portuguese restaurants, whereas in France consumption of raw clams is more common. Popular Portuguese clam dishes are: amêijoas na cataplana, amêijoas à Espanhola and amêijoas à Burhão Pato.

A.5. Legal aspects

The intertidal grounds belong to the State. Collecting shellfish from the natural beds and farming shellfish on special parks was for many years regulated by the decree no. 9637 of May 5th, 1924, but recently a new decree has been issued by the "Ministerias da Marinha, da Economia e da Saúde e Assistência": viz. no. 438/72 of November 7th, 1972, to bring this matter up-to-date. This new regulation is called "Regulamento da Apanha e Exploraçao de Amêijoas e de outros bivalves afins" (Regulation for collecting and exploiting little-neck clams and similar bivalves). It distinguishes:

(a) "Mariscadores", men and women over 14 years old registered as shellfish collectors or as workers in shellfish farms.

(b) Natural beds, where clams may settle naturally.

(c) Licences, authorising the bearer to work on the natural beds or in the parks, and prescribing the utensils which are allowed for this work.

(d) Concessions, giving authorisation to use part of the public property for private farming or storing of clams.

(e) Parks ("viveiros"), plots to be used for growth and fattening of clams.

(f) Depots ("depósitos"), plots for temporary storage of clams.

The licences to be paid are classified as follows:

Mariscadores working on foot, Esc. 50/year; by unpowered boat, Esc. 100/year; by boat equipped with an engine, Esc. 200/year; for parks and depots smaller than 250 m^2, Esc. 250/year; for every further 100 m^2, Esc. 50/year.

A considerable handicap arising from the old regulations was that the maximum area of a park was set at 1 500 m^2 per farmer. This made it impossible to set up an enterprise of any importance, but it was circumvented by licencing more park area fictitiously under the name of other concessionaires. The new regulation gives a maximum area of 5 000 m^2 for each farmer, but in reality there are farmers operating on a considerably larger scale. Mr Francisco Reis Bom exploits two groups of parks: one of 50 000 m^2, controlled by the port of Olhão and one of 35 000 m^2 under the jurisdiction of the port of Tavira. Part of this area is registered under other names, though in reality it is all exploited by the one company.

The majority of parks are in the region of Faro, Olhão and Tavira, and a smaller number of parks are situated near Portimão and Vila Real de Santo Antonio. The parks are duly delineated with pickets and seamarks which should be labelled.

One of the important matters prescribed by the new regulations is that collecting shellfish and working on the parks is only allowed between sunrise and sunset. The use of dredges or suction dredges is not allowed at present, but might soon be considered by the authorities concerned.

The minimum legal sizes of clams for transportation, for home consumption and for exportation, now set at 2 cm, 3 cm and 3.5 cm respectively, are not prescribed by the decree, but can be fixed on an annual basis by the Director-General dos Serviços de Fomento Marítimo, advised by the Portuguese Fisheries Institute (Instituto de Biólogia Marítima). There is no close season for fishing and marketing clams in Portugal.

Official surveillance is carried out by the "cabos de mar", policemen of the ports under which the various stretches of coastline are registered. In addition, the registered "mariscadores" working in the service of the farmers keep watch over the parks, though they are not entitled to arrest trespassers. Theft of clams is a rare occurrence here.

Fig. 90. The small boats of the "mariscadores" in the port of Olhão. (Photo H. Vilela.)

Fig. 91. "Mariscador" of Faro in his boat equipped with outboard engine. (Photo H. Vilela.)

B. SEQUENCE OF OPERATIONS

B.1. Preparing the park

A park, duly delimited by pickets and seamarks, must be smoothed and cleaned before the clams can be planted. Seaweed and stones should be removed. For this one uses a rake, and whenever necessary, a shovel and a hoe. All this work is done by hand.

B.2. Collecting seed clams from the natural beds

The "mariscadores", registered clam fishermen, predominantly male adults, go to the natural beds to collect clams when the tide recedes. Officially there are about 450 registered "mariscadores" in Algarve, but in reality more than a thousand people are engaged in this work. When the bed on which they want to work is nearby, they can go on foot, but usually the "mariscadores" make use of their boat (Figs 90 and 91). The small wooden boats are their personal property. The boats vary in length from 4 to 5 m, in beam from 1 to 1.90 m, in depth from 0.50 to 0.60 m and in weight from 0.65 to 1.36 tons. The boat can be rowed, but most of them are equipped with a sail, some with an outboard engine of about 5 h.p. It takes from 30 to 90 min to reach the working site.

On arrival at the site one can work for about 4 h in succession if the tide goes out far. The collecting of clams on the intertidal grounds is always done on foot, using a peculiar oval-shaped knife, the "faca de mariscar", to dig into the sediments in search for clams. It has a wooden shaft, 12 to 15 cm long and 3 to 3.5 cm in diameter, whereas the oval metal blade is 19 cm long with a maximum breadth of 12 cm.

The seed clams which are collected should officially have a minimum shell-length of 2 cm, but occasionally clams of somewhat smaller size are taken, though never those with a shell-length less than 1.5 cm. The clams are put in wicker hand baskets, 40 cm high and 29 cm in diameter at the top (Figs 92 and 93).

The quantity of clams collected in one tide will differ according to the density of the population, often only amounting to 4 or 5 clams per m^2, a reasonable catch being 3 to 4 kg.

B.3. Transfer of seed clams to the farmer

The mariscadores carrying the collected seed clams in their baskets, deliver them immediately upon their return from the beds to the clam farmers ("viveiristas") who pay them on the spot a price of between Esc. 25 and 35/kg. Mariscadores who possess their own parks will, of course, keep their clams for planting. The farmer grades the clams, when some of them reach con-

Fig. 92. Utensils for collecting little-neck clams: the "faca de mariscar" (oval-shaped knife), and hand basket. (Photo H. Vilela.)

Fig. 93. Collecting little-neck clams for transplanting on intertidal plots. (Photo H. Vilela.)

sumption size, into two categories: marketable clams to be shipped to the market without delay, and seed clams to be planted on the park the next day at low tide.

B.4. Planting the clams on the park

The clams, packed in baskets or sacks, are carried by boat to the parks and spread by hand from the baskets, or with a shovel, after having been deposited in small heaps ("pagelas") on the park. The number of clams planted per m^2 varies with the size of the clams planted and with the area the farmer has at his disposal. A reasonable figure is 500 to 1 000 per m^2. The clams dig themselves in, but it is advisable to come back the next day to see if all clams have dug in, and if not, to help them a little by pushing them by hand in the correct orientation into the sediments. Where the beds appear to be quite hard, it is good practice to loosen the sediments with a rake before planting the clams. For no obvious reason, some clams will crawl out of the sand, and will usually die, exposed on top of the sediments.

B.5. Maintenance of the park

The clams will remain on the parks and grow till they have reached marketable size. To grow from a shell-length of 2 cm to a length of 3—3.5 cm one should expect a sojourn of 2.5 to 3 years on the park; those which were smaller than 2 cm when planted should stay at least 3 years.

The park should be inspected at frequent intervals, and anything which might threaten to smother the clams (for instance seaweeds), should be eliminated promptly. Enemies of the clams, such as starfish, crabs and drills should be controlled as far as is feasible.

B.6. Harvesting the clams

For harvesting marketable clams from the parks the same utensils as described for collecting the seed clams are used. Since, however, the clam population is much denser on a park than on a natural bed, far greater quantities are collected in one tide. The farmer himself, accompanied by members of his family or hired hands, collects the clams.

After being brought to the headquarters of the company the clams harvested from the parks are carefully graded and those considered too small for the market will be returned to the parks, if the smaller ones have not already been sorted out during the harvesting operation on the park itself.

B.7. Marketing the clams

The marketable clams, which have at least 3 cm shell-length for the home markets, or 3.5 cm and over for exportation, have to be shipped without delay. There are usually 80 to 120 marketable clams to a kg. It is not unusual to find clams under 3 cm shell-length in consignments destined for the interior. They are packed in wicker baskets without bails or in sacks, and shipped by train or truck, preferably refrigerated ones for long distance travel and exportation.

Sorting and packing is done at the headquarters ashore, where a small shed is usually reserved for storing clams, utensils and packing material.

If the farmer decides not to sell the harvested clams directly, he takes them to his depot ("depósito"), a section of park specially set aside for stocking clams temporarily. The farmers do not make use of concrete basins or flood basins for temporary storage, but will always keep their clams in a natural situation in the sediments.

C. FARMING RISKS

C.1. Hydrographic conditions

When a heavy downpour leads to a serious drop in the salinity of the water bathing the parks, the clams may die. They cannot withstand a salinity of lower than 21‰.

C.2. Predation

Very small clams are devoured by a variety of predators, but as the size of the clams increases the number of species able to dig them out and to overpower them declines.

Birds such as oyster-catchers *(Haematopus ostralegus)* and curlews *(Numenius arquata)* can be held responsible for serious losses among the smaller clams which live fairly close to the surface, and fish such as *Myliobatis aquila* and *Sparus auratus* account for part of the losses suffered. Some flatfish browse at the protruding tips of the siphons of dug-in clams, but this does not lead to the latter's death, although it may impair their growth.

Serious enemies of clams of all sizes are the starfish *Astropecten bispinosus*, *A. aurantiacus* and *Marthasterias glacialis*, whereas the crabs *Carcinides maenas* and *Eriphia spinifrons* can overpower only the younger specimens. The drills *Murex trunculus* and to some extent also *M. brandaris* kill a few clams, for which empty shells bearing a little round hole are evidence. The farmer tries to kill and remove as many of these predators as possible, but he does not use any type of netting to fence in his parks.

C.3. Diseases and parasites

Though one sometimes observes mysterious mortality among the clams, for instance those individuals which crawl out of the sediments to die, a phenomenon ascribed tentatively to an unknown disease, it is hard to prove which micro-organism is responsible. An abnormal mortality observed in the winter 1971—1972 might have been caused by poisonous plankton ("red-tide"). Sporozoans, for example *Haplosporidium tapetis*, were found in the clams, but there is as yet no indication that they are the cause of mass mortality among clams, though they may lead to retardation of sexual maturation. It is remarkable that the trematode *Bucephalus haimeana*, which is so often found in great numbers in *Venerupis pullastra* at higher latitudes, does not seem to cause any losses on the Algarvian coast.

D. ECONOMIC ASPECTS

D.1. Rents

In the section "Legal Aspects" the new rules for renting a park have been given: Esc. 250/year for an area under 250 m^2, and Esc. 50/year for every further 100 m^2. Moreover, the farmer himself and all his co-workers have to be properly registered as "mariscador": Esc. 50/year on foot, Esc. 100 and 200/year respectively for a boat without, and a boat with motor power.

D.2. Inventory items

The clam farmer has no need to invest much money for his inventory. An expensive item is his boat, equipped with sail or outboard engine. This is a wooden boat of 4 to 5 m length. The utensils, shovels, hoes, rakes, knives, baskets, and some kind of weighing instrument are relatively inexpensive and have a relatively long life. The small wooden shed for sorting and packing clams, and for storing clams, utensils and packing material should also be reckoned among the inventory items.

D.3. Expendable items

The Portuguese clam farmer need not buy much material for his operation. If he operates with a boat equipped with an outboard engine he will have to buy petrol, the amount determined by the trip to and from his parks which require about 1 h sailing on average. The packing material is paid for by the buyer of the clams and when this consists of wicker baskets these may be returned to the shipper.

D.4. Purchase of seed stock

Among the annual expenses of the clam farmer the wages of the personnel and the buying of seed stock largely predominate. If he plants seed clams of 2 cm to harvest them at a shell-length of about 3.5 cm, his individual clams increase their weight from about 2 g to about 10 g in 2.5 to 3 years. To produce 100 tons of marketable clams, the farmer therefore has to purchase 20 tons of seed clams to be planted on the parks. In reality, the tonnage that needs to be purchased is of course much higher, for not all the clams planted will survive. Mr Bom claims that he has to plant five clams to harvest two, a loss of about 60%. This figure does seem rather high and can be considered as pessimistic. Should the losses amount to 60% during the farming period, one would have to buy 50 tons of seed clams (at Esc. 25—35/kg, unsorted) to stock the parks in order to produce 100 tons of marketable clams (with a value of Esc. 38—40/kg).

One could think of reducing the costs of purchasing seed stock by sending members of the personnel to the natural beds to collect seed clams in times of low workload, but the quantities that can be gathered in 1 day are so small (3 to 4 kg) that such a procedure could only cover a small percentage of the tonnage of seed clams required for an enterprise of the scale of that under consideration.

D.5. Labour

The Portuguese clam farmer takes an active part in all the truly artisan work, often assisted by members of the family. The larger firms, like that of Mr Francisco Reis Bom, employ in addition three of four hired hands for various phases of the operations such as preparing the parks for planting clams, maintenance of the parks, harvesting clams from the parks, grading of clams at the headquarters, packing and transporting clams. The wages of the hired hands amount to Esc. 100 to 120/day.

D.6. Sales

The prices paid on the spot for well-graded clams are: *Venerupis decussata*, seed clams, Esc. 40—45/kg; *V. decussata*, marketable clams, Esc. 38—40/kg; *V. aurea*, marketable clams, Esc. 6—8/kg. That *V. decussata* fetches a considerably higher price than other species of the same genus should not be ascribed to a difference in flavour, but to its better preserving qualities.

The retail price for *V. decussata* varies according to the size of the clams and according to the season, and ranges from Esc. 45 to 55/kg, the number of clams counted per kg varying from 80 to 120.

The tonnage shipped by Mr Francisco Reis Bom fluctuates between 100 and 150 tons/year.

E. GOVERNMENTAL ASSISTANCE

The regulations for exploiting clam farms in the coastal waters of Portugal were set up by the Portuguese Government and laid down in the Diário do Governo No. 259 of November 7th, 1972, and were prepared with the assistance of Dr H. Vilela, then Director of the Portuguese Fisheries Institute. Dr Vilela has always shown a keen interest in this matter and can be considered as an authority in the field of European clam farming. Another aspect of the activities of the government is the annual inspection of biological and sanitary conditions on the clam farms ("comissões de vistoria"). This involves authorities of various descriptions, including a member of the Fisheries Institute.

Water samples are analysed bacteriologically on a monthly basis and if they contain over 50 coliform bacteria per 100 cm^3 they are considered to be polluted. Both parks and natural beds bathed in such polluted water can be closed officially, and no new parks will be authorised in such a locality.

Neither the consignments of clams shipped to the interior markets, nor those destined for exportation are covered by certificates guaranteeing their cleanliness. This should be considered in the light of the fact that in Portugal the clams are always cooked before they are consumed.

F. GENERAL IMPRESSION

There are not many cases of clam farming in the world which have passed the experimental phase successfully. Some types of clam farming are to be found in Japan, and for the economically very valuable clam *Mercenaria mercenaria* farming systems in England and on the Atlantic coast of the United States have been tried on the basis of seedlings produced in hatcheries but it is still too early to speak of a commercial success. On the Pacific coast of the United States *Venerupis philippinarum*, once introduced accidentally from Japan, may in due course be farmed commercially.

Farming the little-neck clam *V. decussata* is, however, an established practice of considerable economic importance on the south coast of Portugal. It may seem odd to speak of farming when nothing is done to promote either the production of seed or the growth of the young clams by laying them on grounds which are better suited for growth and fattening than the natural beds, but nevertheless one can safely conclude that the establishment of clam parks has led to a higher production of marketable clams of good quality than would ever have been possible by simply harvesting from the natural beds, since the considerable fishing intensity involved would give few clams a chance to reach a large size.

The clams on the parks live in much denser populations than those on the natural beds, and the parks are located at about the same level as the beds.

Nevertheless, the clams grow well on the parks since it is the water which brings their food. Just as in oyster farming, a great number of individuals can be planted safely on a small area where the tidal currents bring ample water loaded with food. Trouble would be encountered if densely-planted parks covered a very high percentage of the potential farming area.

The total production of little-neck clams from the Algarvian coast is estimated at: *V. decussata*, exportation 400—600 tons; *V. decussata*, home consumption 200 tons; *V. aurea*, home consumption 150—200 tons.

In comparison with these figures the quantities fished from the natural beds in the Tagus, the Sado and in the Lagoon of Aveiro, where no farming takes place, are unimportant.

Figures on the total production of little-neck clams in Portugal given by Vilela (1950) reveal noteworthy fluctuations as follows:

Early in the century the production rose rapidly to about 1 000 tons in 1910, predominantly for home consumption. Soon after, exportation began to rise, first to about 1 000 tons/year, rising to 2 000 tons after 1920 with a remarkable peak of nearly 5 000 tons in 1922. After 1925, the exportation declined again between 1937 and 1949, but for the period 1960—1970 the exportation tonnage fluctuated between 377 and 796 tons, with an average of 586 tons. The figure for home consumption is statistically less reliable, since a great number of small consignments, shipped directly to the consumer, are not recorded.

G. BIBLIOGRAPHY

Anonymous, 1972. Regulamento da apanha e exploraçao de amêijoas e de outros bivalves afins. Diário do Governo I, No. 259, pp. 1598—1603.

Bouxin, H., 1931. Technique d'élevage de deux palourdes comestibles (*Tapes decussatus* L. et *Tapes pullastra* Wood). Rev. Trav. Off. Sci. Tech. Pêches Marit., 9: 101—112.

Vilela, H., 1941. Sobre a biologia da Amêijoa (*Tapes decussatus* L.) 1. Crescimento. 2. Variação sazonal das gónades. Trav. Stat. Biol. Marit. Lisboa, No. 47, 48, 25 pp.

Vilela, H., 1947. Contribução para o estudo das comunidades animals dos parcéis da Ria de Faro. Trav. Stat. Biol. Marit. Lisboa, No. 50, 63 pp.

Vilela, H., 1950. Vida bentónica de *Tapes decussatus* (L.). Trav. Stat. Biol. Marit. Lisboa, No. 13, pp. 1—120. Arquivos do Museu Bocage, 21, Lisboa, 1950.

Vilela, H., 1951. Sporozoaires parasites de la palourde, *Tapes decussatus* (L.). Rev. Fac. Cienc. Univ. Lisboa, Ser. (2) C, 1(2): 379—385.

Vilela, H., 1951. Os moluscos na economia do mar. Bol. Pesca, 31, 39 pp.

Vilela, H., 1956. Exploração dos moluscos bivalves marinhos em Portugal. Realizacoes e Projectes, Lisboa, 1956, 15 pp.

INDEX